The Springer Series on Touch and Haptic Systems is published in collaboration with the EuroHaptics Society. It is focused on publishing new advances and developments in all aspects of haptics. Haptics is a multi-disciplinary field with researchers from Psychology, Physiology, Neurology, Engineering, and Computer Science (amongst others) contributing to a better understanding of the sense of touch, and researching into how to improve and reproduce Haptic interaction artificially in order to simulate real scenarios. The series includes monographs focused on specific topics, edited volumes covering general topics from different perspectives, and selected PhD theses. Books in this series focus on Haptics or Haptic interfaces including:

- Neuroscience
- Perception & Psychophysics
- Haptic Rendering
- Devices and Technology
- Medical & Rehabilitation Applications
- Art & Design Applications
- Collaborative Haptics
- Mutimodal Interaction
- Tactile Display & Tactiles
- Sensing
- Haptic HCI (Interaction, Visualization)

Federica Barontini

Wearable Haptic Devices for Realistic Scenario Applications

Springer

Federica Barontini
School of Engineering
University of Pisa
Pisa, Italy

ISSN 2192-2977 ISSN 2192-2985 (electronic)
Springer Series on Touch and Haptic Systems
ISBN 978-3-031-70538-0 ISBN 978-3-031-70539-7 (eBook)
https://doi.org/10.1007/978-3-031-70539-7

This Springer imprint is published by the registered company Springer Nature Switzerland AG
The registered company address is: Gewerbestrasse 11, 6330 Cham, Switzerland

To the loves of my life, Agata and Enrico,
To my Family, for being my family,
To all the people who makes my heart smile,
To the black sheep.

Preface

The capacity to interact and perceive the environment through our senses is one of the most fascinating human characteristics. Unfortunately, this incredible system can suffer from irreversible injuries that impair our natural sensory-motor capabilities. Among the most severe cases are upper and lower extremity amputations and blindness.

The origin of this work lies in the need to bridge the gap between technological advancements and their practical application in improving human sensory experiences. In recent decades, technological advancements in wearable haptic devices have provided promising alternatives to restore some functions of impaired sensory organs. However, despite these promising results, the effective translation of these technologies into real-world applications remains limited.

One of the primary reasons for this limited acceptance is the insufficient involvement of real users in the conceptual and design phases, with their input typically only sought during the final testing of the device. While this tendency is gradually changing, we are still far from achieving systems designed for, and hence usable by, real people with real needs.

This book proposes different wearable haptic solutions aimed at satisfying users' needs with the ultimate goal of being used in everyday life. The scope encompasses the development, characterization, and validation of haptic feedback devices tailored to user needs. The central idea of this work is to reverse the standard paradigm by adopting a user-centered approach from the initial design phases rather than focusing solely on experimental validation. The purpose is to demonstrate that user-centered design can lead to more effective and widely accepted sensory substitution devices.

The feedback technologies presented here are developed based on user requirements gathered through focus groups, interviews, and multiple experimental sessions. The devices are characterized from both mechanical and psychophysical perspectives, and the effectiveness of the provided stimuli is tested across various applications, from navigation for visually impaired individuals to force feedback for participants with upper limb loss. The plan involves a thorough investigation of existing technologies, user requirements, device development, and comprehensive testing.

Results show significant improvements in terms of efficiency, prosthesis control (especially during grasp actions), and quality of life. These findings confirm that sensory substitution devices could be crucial not only in improving prosthesis acceptance and embodiment but also as valid tools for human-machine interaction training and rehabilitation systems.

The intended audience for this thesis includes researchers, practitioners, and anyone interested in the fields of haptics, rehabilitation robotics, and a user-centered approach. By focusing on the users from the outset, this work aims to create wearable solutions that are effective and truly meet the needs of those who will use them.

Pisa, Italy
June 2024

Federica Barontini

Acknowledgements and Declarations

The work summarized in this book is the result of almost four years of research conducted between 2018 and 2022 at the Research Center "E. Piaggio," at the Department of Information Engineering of the University of Pisa, and at the Soft Robotics for Human Cooperation and Rehabilitation Lab of the Istituto Italiano di Tecnologia in Genoa.

I would like to acknowledge, first and foremost, all the people who accompanied me throughout these years and made it possible to realize many of the ideas presented in this work. The first and deepest acknowledgment goes to my supervisors, Prof. Antonio Bicchi, Prof. Matteo Bianchi, and Prof. Manuel Giuseppe Catalano, for introducing me to the world of research. They always pushed me to do my best and guided me in the right direction whenever I got lost. My deepest gratitude goes to Prof. Kristin Zhao for hosting me during my time abroad at the Mayo Clinic. She allowed me to closely observe the clinical aspect of my work and be part of an amazing rehabilitation team.

A huge acknowledgment to all my friends and colleagues who worked with me, coauthored my manuscripts, and shared this journey with me. To Giuseppe and Federica, Domenico, Alessandro, Vincenzo, Manuel B., The Navacchio core: Mattia and Cristiano, for the laughs, the coffee, the late nights working, the continuous exchanges, and the inspiring discussions that enriched my days. Thank you to the colleagues at the Research Center "E. Piaggio" and the Istituto Italiano di Tecnologia. Without your help and support, none of the ideas presented in this book would have seen the light. You are the best.

Last but not least, I would like to acknowledge my Family. First, a profound thank you to my niece and nephew, Agata and Enrico, for looking at me like a superhero. Thank you for always believing unconditionally in me and for making me feel like the most loved and lucky aunt in the world. I am proud of you, and if I keep going and fighting, it is because of you. To my sister Eleonora, my best friend and greatest supporter, who always stood by me during dark moments and pushed me to see things from a new perspective. A simple acknowledgment to my mom and dad is not enough. I will never be able to repay you for all the effort and sacrifices you have made to allow me to pursue my dreams. If I am the woman I am today, it is because of you, your support, education, and life lessons.

And to all the people I have not mentioned here, thank you for your unwavering presence. As a writer said, "Gli amici sono, come ogni rarità, lussi." Well, I think I am a truly rich person.

Thanks to all of you. You are my Wonderland.

Competing Interests The results illustrated in this book are the fruits of research performed at the Research Center "E. Piaggio" and the Dipartimento di Ingegneria dell'Informazione of the University of Pisa, Soft Robotics for Human Cooperation and Rehabilitation Lab of the Istituto Italiano di Tecnologia in Genova, the Department of Physical Medicine and Rehabilitation of the Mayo Clinic Hospital in Rochester, Minnesota (USA) and the Laboratory for Biomechanics and Biomaterials of the Departments of Orthopedic Surgery, at the Hannover Medical School.

The research leading to these results was possible because of the scholarship offered by the Ph.D. Programme of Dipartimento di Ingegneria dell'Informazione.

These studies were funded by the European Union's Horizon 2020 Research and Innovation Programme: Project SoftPro "Synergy-based Open-source Foundation and Technologies for Prosthetics and RehabilitatiOn" [Grant Agreement ID: 688857], and Project Natural BionicS "Natural Integration of Bionic Limbs via Spinal Interfacing" [Grant Agreement ID: 810346], and by the ERC Proof of Concept Wearable Integrated Soft Haptic Display for Prosthetics (WISH) project [Grant Agreement ID: 101069179].

The results of this work were also supported by the Italian Ministry of Education and Research in the framework of the CrossLab project (Departments of Excellence) and in the framework of PRIN (Programmi di Ricerca Scientifica di Rilevante Interesse Nazionale) 2017, with the project TIGHT: "Tactile InteGration for Humans and arTificial systems" [Grant Agreement ID: 2017SB48FP]. The authors have no conflicts of interest to declare that are relevant to the content of this chapter.

Ethics Approval The experimental studies were conducted in accordance with the principles of the Declaration of Helsinki. All participants enrolled in the various studies provided their consent to participate and to the publication of the results obtained by signing an informed consent form, which informed them of the entire experimental procedure, their rights, and their responsibilities.

The study presented in Chap. 3 was approved by the Local Ethics Committee of Hannover Medical School (7444/18).

The studies presented in Chaps. 4, 7, 8 and 9 were approved by the Bioethics Committee of the University of Pisa.

The study presented in Chap. 5 was approved by the local Ethical Committee of Area Vasta Nord Ovest (CEAVNO, protocol no. 9219).

The study presented in Chap. 6 was approved by the Mayo Clinic Institutional Review Board (IRB).

Pisa, Italy
June 2024

Federica Barontini

Contents

Acronyms

ADLs	Activities of Daily Living
AM-ULA	Activities Measure for Upper Limb Amputee
ASL	American Sign Language
CIs	Confidence Intervals
CNS	Central Nervous System
CPs	Critical Periods
CS	Comparison Stimulus
CUFF	Clenching Upper-limb Force Feedback
DESC	Discrete Event-driven Sensory Feedback
DOFs	Degrees of Freedom
EDC	Extensor Digitorum Communis
EMG	ElectroMyoGraphic
EPT	Extended Psyologic Taction
ETAs	Electronic Travel Aids
FDS	Flexor Digitorum Superficialis
FEPAP	Federation of European Producers of Abrasive Products
GLMM	Generalized Linear Mixed Model
GM	Grey Matter
HCI	Human-Computer Interaction
HP	Haptuator Planar
HRI	Human-Robot Interaction
IMUs	Inertial Measurement Units
JND	Just Noticeable Difference
MM	Modality Matching
PLP	Phantom Limb Pain
PSE	Point of Subjective Equality
Psoc	Programmable System-on-Chip
PWM	Pulse Width Modulation
RMSE	Root Mean Square Error
RS	Reference Stimulus
SH	SoftHand

SM	Somatotopic Matching
SPH	SoftHand Pro
SSC	Somatosensory Cortex
SSDs	Sensory Substitution Devices
SUS	System Usability Scale
TR	Target Reinnervation
UIC	Unione Italiana Ciechi ed Ipovedenti
UIs	User Interfaces
VI	Visually Impaired
VIBES	Vibro-Inertial Bionic Enhancement System
WISH	Wearable Integrated Soft Haptic
wOCHD	Wearable Outer-Covering Haptic Display

Chapter 1
Introduction

Abstract How do you perceive the world? When someone asks this question, the simplest and most logical answer might be "through the information coming from our senses". Since ancient times, numerous theories have emerged regarding the role of the senses and the perception of reality. It should come as no surprise how traumatic the loss of one or more sensory modalities can be for our life experience. The aim of this book is to assist those who have suffered such trauma by leveraging one of our primary intrinsic recovery mechanisms for sensory loss: neural plasticity, through the use of wearable haptic systems. This first chapter aims to present a concise overview of the theory of perception through the ages, explore how our bodies respond to sensory loss via neural plasticity, and discuss the related haptic interfaces designed to address these issues. Special attention is given to two disabilities intensively explored in the state of the art: blindness and upper limb loss.

1.1 How Do We Perceive the World?

This paragraph provides an overview of some of the main theories of perception and the study of the senses across various disciplines, from philosophy and psychology to modern neuroscience. The discussion traces the evolution of thought from ancient philosophical debates to contemporary scientific theories, highlighting the contributions and disagreements among different schools of thought.

1.1.1 Understanding Human Senses and Perception: A Historical Overview

How do you perceive the world? When someone asks us this question, the most simple and logical answer could be "through the information coming from our senses".

How can we define *Perception*? The term comes from the Latin *perceptio, percipio* and identifies the organization, identification and interpretation of the sensory

F. Barontini, *Wearable Haptic Devices for Realistic Scenario Applications*,
Springer Series on Touch and Haptic Systems,
https://doi.org/10.1007/978-3-031-70539-7_1

information in order to represent and understand the environment (Schacter et al. 2019). Thanks to our senses, we have a specific idea of the world.

The theory of perception has been affected by many "schools" of thought. Each has its central theoretical viewpoint and its own particular set of methodological techniques. These include psychophysicists, Gestaltists, functionalists, structuralists, transactionalists, sensory physiologists, sensory-tonic theorists, "new look" psychologists, efferent theorists, cognitive theorists, and information processors, to name a few. To give the reader the first deal with the *perceptual problem*, which is the issue of the correspondence (or noncorrespondence) between our internal representation of the environment in consciousness and the objectively measured external physical situation, we need to start the discussion with some philosophical consideration.

The Athenian philosopher Plato (428–348 B.C.) argued that we perceive objects through the senses but with the mind. The basic notion is that the sensory inputs are variable and inaccurate, and they can provide only an imperfect copy of the objects. The reason or intellect has the job of correcting these inaccuracies to provide us with a true and correct picture of the world.

A completely different point of view was asserted by Protagoras (480–411 B.C.), the founder of the doctrine known as Sensism. He claimed that "Man is nothing but a bundle of sensations." This idea was supported by the school of Aristippus (435–366 B.C.), who declared that the senses are inherently precise and thus responsible for our detailed view of the environment. Hence, there should always be a good correspondence between perception and reality. If there are any distortions, it is the mind or judgmental capacities that are limited and responsible for the discrepancies (Boring 1942). Only with Aristotele (384–322 B.C.), we could find a compromise between these two positions. According to the philosopher, sensation (*aisthesis*) is the first level of knowledge because the senses are passive receptors of the word (Everson 1997). Some perceptual qualities are immediately and accurate perceived by the senses. Aristotle stated that what is experienced during perception is active concerning the senses themselves: it is the nose that receives the smell, it is the eye that receives the colour, it is the ear that receives the sound, and so on. The sense organ is pure perceptual potential, activated when the sensible quality presents itself; therefore, in the act of reception, the sense is presented as passive as it "becomes assimilable" (Johansen 1997).

However, other qualities, such as movement, number, figural qualities, and magnitude, are not the exclusive property of any sense but are common to all. According to Aristotelian doctrine, these qualities require intellectual meditation to assure the accuracy of representation.

Transitioning to a later philosophical era, the sensist doctrine reached its height with John Locke (1632–1704), one of the foremost exponents of Empiricism, who proposed a clean-state view of the brain (Locke 1948). In 1689, Locke suggested that we are born without any preconceptions or innate ideas (*tabula rasa*), and that experience ultimately shape the brain thus, nurture determines who we are. This notion is central to Locke's empiricism, which emphasizes individual's ability to author their destiny. Locke's view of perceptual experiences was further developed in the twentieth century by Jean Piaget (1896–1980) when, in his 1969 book *Mechanisms of Perception* (Piaget 2013), Piaget hypothesized that there is no difference between perception and intelligence.

By the mid nineteenth-century, the notion of associating information was the law governing brain functions started seeking by philosophers, psychologists, early physiologists, neurosurgeons, and the first neuroscientists. The physiologist Helmholtz (1821–1894) introduced mechanisms by which a sensation became a perception. In his theory of perception (1848–1868), he argued that the mind makes a series of mental adjustments, *"unconscious inferences"*, to construct a coherent picture of its experience (Heidegren 2018). Helmholtz's theory suggests that individuals become conscious of an existing object in the environment using inferences based on their experience. He described the basic event as an "impulse" and sensation as an "aggregate". Each basic event corresponds to an "immediate perception" connected with the impulse and the sensation aggregate. These events are inductive inferences that generate a representation. Shared cultural and environmental factors mean there is generally good agreement on the nature of objects and relationships in the world. However, individual differences in personal histories can lead to different perceptions among people given the same stimulation. The inclusion of experience as a factor in shaping the final percept means that perception has a developmental aspect and will differ as a function of the individual's age.

Over the years, Helmholtz's viewpoints evolved into the "New Look" theories of perception which allowed for a broader spectrum of experiential and inferential contributions. These theories then developed into information-processing theories and, finally, the modern conception of cognition. There is, however, one theoretical approach that harkens back to the early sensist approaches and includes a relatively emphatic denial of contributions from reason or intellect.

At the beginning of the twentieth century, Gibson (1904–1979) introduced the theory of ecological psychology (Gibson 2014). In its very origins, he aimed to provide an innovative perspective for understanding perception and perceptual learning overcoming traditional psychological dichotomies such as perception/action, organism/environment, subjective/objective, and mind/body. According to Gibson, perception can be explained by analyzing the structure of an organism's environment (the ecology of the organism), rather than the brain's structure. The organization of the environment conveys the information on which the perception is based. Several other significant findings were achieved over the years; however, these will be not reviewed in this Introduction, leaving space for a discussion on modern theories on brain plasticity in the following sections.

1.2 The Role of Neuroplasticity and Recovery from Sensory Impairment

Neuroplasticity, also known as brain plasticity, is the remarkable ability of the central nervous system to adapt to new experiences and stimuli. This section explores the foundational concepts of neuroplasticity, tracing its historical development and highlighting key factors that contribute to the brain's capacity to reorganize itself. Understanding these mechanisms is crucial for comprehending how the brain responds to sensory loss and other changes, ultimately aiding in the development of effective rehabilitation strategies.

1.2.1 The Role of Neuroplasticity: How Our Brain Adapts Itself to New Circumstances

From the neuroscientific perspective, the capacity of the central nervous system to adapt to stimuli and experience is called Neuroplasticity or Brain Plasticity.

Brain Plasticity is "*...the adaptive capacities of the central nervous system—its ability to modify its own structural organization and functioning*" (Batch-y Rita 1988). In other terms, neuroplasticity could be seen as the capacity of the brain to modify, change, and adapt both structure and function throughout life and in response to experience.

Before the nineteenth century, philosophers mainly contemplated the brain, so it was only in the late 1800s and early 1900s that the foundations were laid for modern neuroscience. In the last decade of that century, several scientists made critical contributions to our modern understanding of synaptic plasticity but was the Spanish Neuroanatomist Santiago Ramon y Cajal first defined the neuron as the nervous system's anatomical, physiological, genetic, and metabolic unit in his Neuron Doctrine. With his hypothesis about cerebral gymnastics, Cajal suggests that the brain's capacity could be augmented by increasing the number of connections. In 1893, the Italian Neuropsychiatrist Eugenio Tazi, supporting Cajal's theory, proposed that repetitive activity in a neuronal pathway could produce hypertrophy through specific learning or practice, thereby reinforcing the already existing connection. To deeply understand how brain plasticity works, it is necessary to analyze the cortical factors that contribute more to neuroplasticity: critical periods, plasticity inhibitors and the neuromodular system. The Critical Periods (CPs) are a particular class of "*sensitive periods*". This term identifies the effects of experience on the brain that are usually strong during a limited period of development (Knudsen 2004). They represent periods in development during which specific capacities are modelled or altered by experience. The second factor is the plasticity inhibitors, cells that have the role of limiting the plasticity in the aged brain. The last is the neuromodular systems, a group of neurons which have powerful effects on cognitive behaviour. The neuromodular systems include neurotransmitters (i.e. noradrenergic, serotonergic, dopaminergic, and cholinergic systems), which are the basis for many of the high cognitive functions, attention, emotion and decision-making. The interaction between the neuromodular system and brain areas determines specific behaviour (for a detailed description of the neuromodular system, please refer to Avery and Krichmar 2017). For almost a century, the traditional view of the plastic nature of the brain was persuade that plasticity was limited to short epochs during early development. It is only recently that researchers started to have a complete understanding of the different regulators that modulate and underlie plasticity.

A recent study on the auditory system of young and old rats de Villers-Sidani et al. (2008) suggests how the elements that regulate plasticity change during the lifespan and not only around the CPs. Moreover, multiple studies on CPs demonstrated the importance of sensory experience for normal neuronal development and sensory map acquisition. However, the quantity and quality of the experience have different effects on CP in terms of duration and outcome. What is clear is that the excessive presence of a specific stimulus during CP manifests in incorporation into the sensory map. A study on musicians analyzed the plastic changes in the human brain. It is not easy to see the plastic changes in the human brain; in fact, most of the studies were conducted on the animal, but the brain of a musician represents an ideal model for the complexity of the stimulus (i.e. the music) and their exposure to this stimulus (Münte 2002). The study confirms that what emerged in studies with rats that undergoes the brain to exceed stimulation for a specific amount of time, triggers morphological changes in the brain.

Voss et al. (2017) analyzed how interaction between the critical period, neuromodular system, plasticity inhibitor and individual factors like age, sex, and sensory experience produce variability in the plastic processes. Their study demonstrated that multiple factors could potentiate the experience-dependent plasticity throughout life. These findings permit to claim that the plasticity mechanism and regulator can also be manipulated in the mature brain by acting on neuromodulators and precise sensory input.

Moreover, in their review Gu and Kanai (2014) identify that genetic and environmental influences largely determine individuals variation in brain structure. They analyzed twin studies and demonstrated that genetic variants contributes to individual differences in distinct brain areas. On the other hand, to investigate the environmental influences, the authors compared a cohort of participants (expert and non-expert in particular fields), and they established that there is a connection between environmental influence and morphological changes in brain structure. For example, a professional dancer or experienced meditator (Hölzel 2008; Hänggi et al. 2010) has a greater grey matter (GM) in a specific part of the brain.

Plasticity is also a phenomenon that aids brain recovery after damage produced by events (i.e. a stroke or traumatic injuries), but also when the brain is deprived of one sensory modality. When we lose one of our senses, our body responds by activating brain plasticity, engaging in sensory substitution as a spontaneous reaction mechanism. This natural adaptability allows the brain to reorganize and compensate for the lost sensory input by enhancing the remaining senses. Next section delves into sensory substitution, particularly in the contexts of blindness and limb loss. These two forms of sensory deprivation have been extensively studied in the field of neuroplasticity, providing a wealth of research and practical insights. Focusing on blindness and limb loss is crucial because these conditions present unique challenges and opportunities for understanding how the brain adapts to significant sensory loss.

1.2.2 How Do We Recover from Sensory Impairment?

As discussed in the previous section, the central nervous system's (CNS) most remarkable capability is adapting to the experiences. However, what happens when people lose one sense, such as the touch, after a major limb loss or the sight? How does the brain react?

Sensory deprivation involves massive alterations of the brain structure and function, behavioural outcomes, and neural interaction. It has been observed that, after damage, the brain activates several mechanisms to face a new reality. More specifically, these mechanism include:

- **Sprouting**: This is the process whereby a neuron generates additional branches (outgrowths) to establish new links between existing neurons. The sprouting could be adaptive or maladaptive (i.e. the phantom limb pain after amputation is an example of maladaptive sprouting.)
- **Denervation Supersensitivity**: This results from a permanent increase in neuronal responsiveness to diminished input. For example, the receptor site might become more sensitive to a neurotransmitter or increase the number of receptors. Behaviour compensation involves using a group of muscles and/or cognitive strategy to replace the missing part.
- **Unmasking**: This process changes excitability or inhibition through an external process. Neurologically, it is the activation of neural pathways or synapses that are inhibited in the normal state but become active after brain damage. Unmasking is essential for functional recovery, but like sprouting, it can also have a negative effect (e.g. the Babinsky reflex).

In the case of blindness, the general loss hypothesis (Rauschecker 1995) involved a "compensatory plasticity" mechanism. According to this theory, early sensory deprivation and the lack of visual experience during postnatal development of the brain induces a generalized degradation of sensory functions. So blind individuals would be impaired in the calibration of their other senses and therefore be unable to form a map of the surrounding space. However in the last decade, several studies have contrasted the hypothesis of compensatory adaptation. Voss et al. (2004) and Leclerc et al. (2000) demonstrated that the brain deprived of one sensory modality shows massive reorganization and adapts to sensory loss by developing superior abilities and recruiting deafferented cortex to process information from preserved modalities. In the case of blindness, the absence of visual information in the occipital cortex promotes the activation of different pathways resulting in improved activation of hearing and touch. Especially for navigation tasks, blind individuals make extensive use of auditory information. The study of Rauschecker (1995) showed that visual deprivation does not prevent the blind from forming a spatial map of their environment and, in fact can perform better than the sighted.

In the case of amputation of a major limb, the part of the brain involved is the contralateral hemisphere. More specifically, the Somatosensory Cortex (SSC) is then responsible for interpreting the word throughout the sense of touch. In the case of amputation, only this part of the brain slows down its functions, but sometimes there is a hyperactivation of the area. It is the case of the Phantom Limb Pain (PLP), where the creation of a new connection between the neurons next to the SSC area, and the inactive neurons generates a new neural path, causing an hyperactivation of SSC. In Preißler et al. (2013), the authors studied on the changing of volume in the Grey Matter (GM) after amputation. The study investigate the grey matter of 21 upper limb loss participants and 14 able-bodied subjects. The results demonstrated that the amputation is associated with a reduction of the grey matter in areas in the motor cortex representing the amputated limb. Additionally, the participants with limb loss show an increase of the grey matter in the region of the visual stream. Results indicate that the patients may have an enhanced need for visual feedback and control to compensate for the lack of sensory feedback of the missing limb.

According to Bach-y-Rita and Kercel (2003), a person who suffers from blind loss does not lose the capacity to see. They lose the peripheral sensory system (the retina) but retain a central visual mechanism. Similarly, a person who becomes deaf or loses a major limb loses the peripheral structure through which they relate/receive information from the environment, maintaining the neuronal structure. Understanding how the plasticity mechanism works after brain damage is the basis for actuating powerful and effective rehabilitation plans. Brain modifications are triggered by sensory deprivation and subsequently by the training of the other senses. Sensory Substitution Devices (SSDs), for example, enable individual to "perceive" visual or tactile information through alternative sensory channels.

In the next section, we provide an overview of the principal sensory substitution system to understand the current state of the art, analyze their limitations, and discuss the main achievements reached so far.

1.3 Enhancing Sensory Substitution Using Haptic Interfaces: The Use Cases of Blind People and People with Limb Loss

Haptic interfaces have become a cornerstone in the development of systems designed to provide sensory feedback, particularly for individuals with sensory impairments. These devices play a crucial role in compensating for lost sensory modalities by providing alternative forms of feedback.

1.3.1 Introduction on Haptic Interface

Touch is an exceptionally powerful source of information, with a wide array of data from the external environment continuously processed by receptors distributed across the human body (Katz and Krueger 2013). Given that the skin is the largest organ, accounting for approximately 15% of body mass (Biswas and Visell 2019), touch is one of the most developed senses and is extensively used in sensory substitution applications.

In recent years, substantial efforts have been directed toward the development of human-robot interfaces that effectively convey haptic cues to users. These haptic interfaces are mechatronic devices equipped with sensors, actuators, and control mechanisms designed to sense touch-related information from interactions with the environment and deliver it to users through haptic stimuli. Their primary aim is to enhance user performance and immersion in various fields, including teleoperation (Ajoudani et al. 2014; Pacchierotti et al. 2015), medical robotics (El Rassi and El Rassi 2020), assistive and rehabilitation robotics (Bortone et al. 2017), guidance (Kuang et al. 2022), virtual reality and entertainment (Dangxiao et al. 2019).

Given the substantial challenges and extensive research dedicated to addressing the needs of individuals with upper limb loss and visual impairments, these two areas have become focal points in the development of haptic feedback systems. These disabilities present unique and compelling opportunities to explore and expand the potential of haptic technologies in restoring lost sensory functions and improving the quality of life for affected individuals.

Sensory feedback in prosthetics has been a highly discussed topic in the past decade, with several reviews published investigating both invasive (Pasluosta et al. 2018) and non-invasive solutions (Svensson et al. 2017), as well as describing devices based on the technology used and the methods employed to elicit tactile sensation (Antfolk et al. 2013). A different perspective is presented in the review by Sensinger and Dosen (2020), where the focus is not on the technologies used or the stimuli conveyed but rather on placing feedback within the broader framework of human motor control and the fundamental role that feedback plays in prosthesis control.

According to Kim and Colgate (2012) and Schofield et al. (2014), haptic feedback can be divided into three main categories:

- **Somatotopic Matching Feedback**: Cues are delivered to the same place where they would be naturally felt. Compared with substitution or modality-matching methods, somatotopically matched feedback may reduce the cognitive burden placed on the user. As a result, the user may require less training and conscious attention to interpret feedback signals. Somatotopic-matching techniques have been investigated with direct neural stimulation such as targeted reinnervation (Kuiken et al. 2007) and peripheral nerve stimulation (Oddo et al. 2016; Tan et al. 2014; Raspopovic et al. 2014).
- **Modality Matching Feedback**: Artificial cues that are provided using the same sensory modality as the natural sensation. This includes non-invasive techniques to restore the haptic channel, i.e. pressure sensed in a prosthetic hand is delivered

as pressure on the arm. Examples include mechanotactile solutions that convey normal and tangential forces by squeezing the arm or stretching the skin (Casini et al. 2015; Colella et al. 2019; Rossi et al. 2018), and rotational skin stretch devices that provide kinesthetic feedback from a prosthetic device (Bark et al. 2010; Aggravi et al. 2018).
- **Sensory Substitution Feedback**: This feedback does not rely on the physiological representations of what the missing sense would experience, but instead requires the user to learn how to interpret the artificial stimuli in relation to the actual haptic information. Example include vibrotactile feedback where the prosthesis can convey information to the users regarding the first contact and the texture of an object, by modulating the frequency of vibration (Fani et al. 2019; Cipriani et al. 2011; Stepp and Matsuoka 2011), and the electrocutaneous stimulation (Wang et al. 1995).

While significant advancements have been made in upper limb prosthetics, it is equally important to consider the progress in sensory feedback systems for visually impaired individuals. Various solutions have been investigated to enhance the autonomy of blind people, particularly in navigation. Several comprehensive reviews have explored different navigation tools and techniques used by visually impaired or blind individuals. For instance, Tapu et al. (2014) conducted a review on wearable devices designed to assist visually impaired users in outdoor environments. Silva (2017) examined vision assistive methodologies for indoor positioning and navigation. The study conducted by Dakopoulos and Bourbakis (2009) discusses several Electronic Travel Aids (ETAs) that convey spatial information about position or obstacles through sensory substitution. These ETAs can be either portable or wearable. Wearable solutions are particularly advantageous as they can be positioned on multiple parts of the body, used alongside traditional aids such as the white cane, and allow users to keep their hands free for interacting with the environment. Hojjat (2018) provided an overview of existing indoor navigation solutions for blind and visually impaired individuals, while Bineeth (2020) offered a comprehensive review of navigation solutions that are effective in both indoor and outdoor environments.

Despite all technological advancements, the acceptance and usage of sensory feedback in users' daily lives or clinical use remain incomplete. Studying the state of the art alone is not sufficient for creating an efficient device, as they often turn out to be too cumbersome for prolonged use, and the mental effort required to interpret the conveyed stimuli can be substantial.

According to Sensinger and Dosen (2020), part of the cause can be attributed to insufficient knowledge and understanding about the fundamental role of feedback in prosthesis control. Another possible explanation is the insufficient involvement of users during the initial development phases. Typically, once the problem to be solved is identified, the focus shifts to developing a device that can reliably replicate the missing sense from an engineering perspective. However, this approach often overlooks the user and their needs. Users are frequently involved only in the final stages of design when system validation is required.

This work proposes to "invert" this cycle by incorporating the user and their needs from the early stages of design, creating a device that meets their requirements while also being effective in sensory reproduction. The devices presented in the following chapters are designed to be non-invasive, satisfying the paradigms of somatotopic and modality matching, and providing tactile stimuli capable of skin-stretch and squeeze/pressure to reproduce proprioception and force feedback.

1.4 Contents of the Monograph

In this book, I intend to propose a new working paradigm that attempts to change the "traditional" approach: moving from a technology-driven to a user needs-driven development. In Chap. 2, I present the interviews conducted with blind and upper limb loss participants regarding their needs and requirements. Chapter 3 discusses a comparison between different feedback modalities to understand which type of information limb loss participants preferred. Lessons learned during this experience are exploited in Chap. 4, where the redesigned version of the Clenching Upper Limb Force Feedback (CUFF) is presented. The proposed device was miniaturized, and in accordance with the users' requirements, the control strategy was improved to elicit a more powerful force stimulus. The device was also characterized and validated through experiments involving able-bodied participants.

Chapter 5 presents the application of the CUFF device with blind users. Specifically, it demonstrates how the integration of the device in a navigation system can enable blind users to walk safely in unknown environments. Chapter 6 presents an experimental campaign conducted at the Mayo Clinic Hospital, Rochester, MN (USA), with upper limb loss participants. The goal of the study was to investigate whether the feedback from CUFF could improve the modulation of force during grasping actions. Building on the lessons learned in previous chapters, Chap. 7 proposes a novel force feedback device fully integrated within the socket of the prosthesis. A detailed characterization of the device from physical and psychophysical perspectives is provided, and preliminary tests with a pilot limb loss user are presented.

Inspired by the results obtained with the integrated force feedback, Chap. 8 presents the integration of a vibrotactile device within the prosthetic socket, capable of transmitting high-frequency contact information such as surface roughness and initial contact from interactions with external objects. The device was characterized from a psychophysical point of view, and its efficacy was tested with a pilot subject in multiple manipulation tasks.

Finally, Chap. 9 explores the application of the CUFF technology in Human-Robot Interaction scenarios, testing the device during manipulation tasks with a tele-robotic platform.

Each chapter is written to be as self-contained as possible, so it can be read either as part of the book or as a stand-alone. Furthermore, other chapters are referred to in the text when needed, to ensure easier navigation through the manuscript.

1.5 Novelties

In this book, I present several contributions to the State of the Art across multiple application fields, ranging from upper limb prosthesis applications to the development of navigation systems for visually impaired users. This work aims to create feedback devices that serve as effective solutions for real users in everyday life. With this goal in mind, I adopted a User-Centered approach in developing all the wearable haptic solutions discussed in this thesis, actively involving users from the design phase onwards.

One of the key novelties of this work is the extensive involvement of users throughout the development process. This was achieved through conducting interviews and participating in focus groups to understand user requirements and priorities. By integrating user feedback early and continuously, I was able to ensure that the devices met their real needs and preferences.

A significant innovation is the development of a novel force feedback device that is completely integrated inside a prosthetic socket. This device can convey pressure information to the user's residual limb, corresponding to the force exerted by the robotic hand during grasp actions. This integration ensures that the feedback is intuitive and directly relevant to the user's interaction with the environment.

Another major contribution is the application of haptic feedback for navigation systems. By utilizing tactile device able to convey directional information, I developed a system that allows visually impaired individuals to navigate unknown environments safely. This system has been validated through extensive testing with blind participants, demonstrating its effectiveness in improving mobility and safety.

The devices developed were subjected to rigorous testing and validation through multiple experimental campaigns involving able-bodied individuals, blind participants, and those with upper limb loss. This comprehensive approach ensured that the effectiveness of the stimuli provided by the devices was thoroughly evaluated across different user groups and scenarios.

The results obtained underscore the importance and necessity of feedback devices in restoring missing sensations and enhancing prosthesis embodiment. Additionally, these devices show significant potential as powerful tools in rehabilitation and training scenarios, providing users with the feedback needed to improve their interaction with prosthetic devices and navigate their environments more effectively.

In summary, this work represents a significant advancement in the field of wearable haptic devices, emphasizing the importance of a User-Centered design approach. The lessons learned and the devices developed contribute to the broader goal of creating more intuitive, effective, and user-friendly solutions for individuals with sensory impairments.

References

Aggravi M, Pausé F, Giordano PR, Pacchierotti C (2018) Design and evaluation of a wearable haptic device for skin stretch, pressure, and vibrotactile stimuli. IEEE Robot Autom Lett 3(3):2166–2173
Ajoudani A, Godfrey SB, Bianchi M, Catalano MG, Grioli G, Tsagarakis N, Bicchi A (2014) Exploring teleimpedance and tactile feedback for intuitive control of the pisa/iit softhand. IEEE Trans Haptics 7(2):203–215
Antfolk C, D'alonzo M, Rosén B, Lundborg G, Sebelius F, Cipriani C (2013) Sensory feedback in upper limb prosthetics. Expert Rev Med Dev 10(1):45–54
Avery MC, Krichmar JL (2017) Neuromodulatory systems and their interactions: a review of models, theories, and experiments. Front Neural Circuits 11:108
Bach-y-Rita P, Kercel SW (2003) Sensory substitution and the human-machine interface. Trends Cognit Sci 7(12):541–546
Bark K, Wheeler J, Shull P, Savall J, Cutkosky M (2010) Rotational skin stretch feedback: A wearable haptic display for motion. IEEE Trans Hapt 3(3):166–176
Batch-y Rita P (1988) Brain Plasticity. Goodgold, J. ed
Bineeth K (2020) Tools and technologies for blind and visually impaired navigation support. a review. IETE Techn Rev 1
Biswas S, Visell Y (2019) Emerging material technologies for haptics. Adv Mater Technol 4(4):1900042
Boring EG (1942) Sensation and perception in the history of experimental psychology
Bortone I, Leonardis D, Solazzi M, Procopio C, Crecchi A, Briscese L, Andre P, Bonfiglio L, Frisoli A (2017) Serious game and wearable haptic devices for neuro motor rehabilitation of children with cerebral palsy. In: Converging clinical and engineering research on neurorehabilitation II: Proceedings of the 3rd International Conference on NeuroRehabilitation (ICNR2016), October 18–21, 2016, Segovia, Spain, pp 443–447. Springer International Publishing
Casini S, Morvidoni M, Bianchi M, Catalano M, Grioli G, Bicchi A (2015) Design and realization of the cuff-clenching upper-limb force feedback wearable device for distributed mechano-tactile stimulation of normal and tangential skin forces. In: 2015 IEEE/RSJ international conference on Intelligent Robots and Systems (IROS), pp 1186–1193. IEEE
Cipriani C, D'Alonzo M, Carrozza MC (2011) A miniature vibrotactile sensory substitution device for multifingered hand prosthetics. IEEE Trans Biomed Eng 59(2):400–408
Colella N, Bianchi M, Grioli G, Bicchi A, Catalano MG (2019) A novel skin-stretch haptic device for intuitive control of robotic prostheses and avatars. IEEE Robot Autom Lett 4(2):1572–1579
Dakopoulos D, Bourbakis NG (2009) Wearable obstacle avoidance electronic travel aids for blind: a survey. IEEE Trans Syst Man Cybern Part C (Appl Rev) 40(1):25–35
Dangxiao W, Yuan GUO, Shiyi LIU, Zhang Y, Weiliang X, Jing X (2019) Haptic display for virtual reality: progress and challenges. Virtual Real & Intell Hardware 1(2):136–162
de Villers-Sidani E, Simpson KL, Lu YF, Lin RC, Merzenich MM (2008) Manipulating critical period closure across different sectors of the primary auditory cortex. Nat Neurosci 11(8):957–965
El Rassi I, El Rassi JM (2020) A review of haptic feedback in tele-operated robotic surgery. J Med Eng & Technol 44(5):247–254
Everson S (1997) Aristotle on perception. OUP Oxford
Fani S, Di Blasio K, Bianchi M, Catalano MG, Grioli G, Bicchi A (2019) Relaying the high-frequency contents of tactile feedback to robotic prosthesis users: design, filtering, implementation, and validation. IEEE Robot Autom Lett 4(2):926–933
Gibson JJ (2014) The ecological approach to visual perception, Classic edition. Psychology Press
Gu J, Kanai R (2014) What contributes to individual differences in brain structure? Front Hum Neurosci 8:262
Hänggi J, Koeneke S, Bezzola L, Jäncke L (2010) Structural neuroplasticity in the sensorimotor network of professional female ballet dancers. Hum Brain Mapping 31(8):1196–1206
Heidegren CG (2018) Three positivist disputes in the 1960s. J Hist Anal Philos 6(8):1–18

Hojjat A (2018) Enhanced navigation systems in gps denied environments for visually impaired people: a survey. arXiv:1803.05987
Hölzel BK, Ott U, Gard T, Hempel H, Weygandt M, Morgen K, Vaitl D (2008) Investigation of mindfulness meditation practitioners with voxel-based morphometry. Soc Cognit Affect Neurosci 3(1):55–61
Johansen TK (1997) Aristotle on the sense-organs
Katz D, Krueger LE (2013) The world of touch. Psychology Press
Kim K, Colgate JE (2012) Haptic feedback enhances grip force control of sEMG-controlled prosthetic hands in targeted reinnervation amputees. IEEE Trans Neural Syst Rehabil Eng 20(6):798–805
Knudsen EI (2004) Sensitive periods in the development of the brain and behavior. J Cognit Neurosci 16(8):1412–1425
Kuang L, Aggravi M, Giordano PR, Pacchierotti C (2022) Wearable cutaneous device for applying position/location haptic feedback in navigation applications. In: 2022 IEEE Haptics Symposium (HAPTICS), pp 1–6. IEEE
Kuiken TA, Marasco PD, Lock BA, Harden RN, Dewald JP (2007) Redirection of cutaneous sensation from the hand to the chest skin of human amputees with targeted reinnervation. Proc Natl Acad Sci 104(50):20061–20066
Leclerc C, Saint-Amour D, Lavoie ME, Lassonde M, Lepore F (2000) Brain functional reorganization in early blind humans revealed by auditory event-related potentials. Neuroreport 11(3):545–550
Locke J (1948) An essay concerning human understanding 1690
Münte TF, Altenmüller E, Jäncke L (2002) The musician's brain as a model of neuroplasticity. Nat Rev Neurosci 3(6):473–478
Oddo CM, Raspopovic S, Artoni F, Mazzoni A, Spigler G, Petrini F, Giambattistelli F, Vecchio F, Miraglia F, Zollo L, Di Pino G, Camboni D, Carrozza MC, Guglielmelli E, Rossini PM, Faraguna U, Micera S (2016) Intraneural stimulation elicits discrimination of textural features by artificial fingertip in intact and amputee humans. elife 5:e09148
Pacchierotti C, Meli L, Chinello F, Malvezzi M, Prattichizzo D (2015) Cutaneous haptic feedback to ensure the stability of robotic teleoperation systems. Int J Robot Res 34(14):1773–1787
Pasluosta C, Kiele P, Stieglitz T (2018) Paradigms for restoration of somatosensory feedback via stimulation of the peripheral nervous system. Clin Neurophysiol 129(4):851–862
Piaget J (2013) The mechanisms of perception. Routledge
Preißler S, Feiler J, Dietrich C, Hofmann GO, Miltner WH, Weiss T (2013) Gray matter changes following limb amputation with high and low intensities of phantom limb pain. Cerebral Cortex 23(5):1038–1048
Raspopovic S, Capogrosso M, Petrini FM, Bonizzato M, Rigosa J, Di Pino G, Capaneto J, Controzzi M, Boretius T, Fernandez E, Granata G, Oddo CM, Citi L, Ciancio AL, Cipriani C, Carrozza MC, Jensen W, Guglielmelli E, Stieglitz T, Rossini PM, Micera S (2014) Restoring natural sensory feedback in real-time bidirectional hand prostheses. Sci Transl Med 6(222):222ra19-222ra19
Rauschecker JP (1995) Compensatory plasticity and sensory substitution in the cerebral cortex. Trends Neurosci 18(1):36–43
Rossi M, Bianchi M, Battaglia E, Catalano MG, Bicchi A (2018) HapPro: a wearable haptic device for proprioceptive feedback. IEEE Trans Biomed Eng 66(1):138–149
Schacter D, Gilbert D, Wegner D, Hood B (2019) Psychology, 3rd European edn. Red Globe Press
Schofield JS, Evans KR, Carey JP, Hebert JS (2014) Applications of sensory feedback in motorized upper extremity prosthesis: a review. Expert Rev Med Dev 11(5):499–511
Sensinger JW, Dosen S (2020) A review of sensory feedback in upper-limb prostheses from the perspective of human motor control. Front Neurosci 14:345
Silva CS, Wimalaratne P (2017) State-of-art-in-indoor navigation and positioning of visually impaired and blind. In: 2017 seventeenth international conference on advances in ICT for emerging regions (ICTer). pp 1–6. IEEE

Stepp CE, Matsuoka Y (2011) Vibrotactile sensory substitution for object manipulation: amplitude versus pulse train frequency modulation. IEEE Trans Neural Syst Rehabil Eng 20(1):31–37

Svensson P, Wijk U, Björkman A, Antfolk C (2017) A review of invasive and non-invasive sensory feedback in upper limb prostheses. Expert Rev Med Dev 14(6):439–447

Tan DW, Schiefer MA, Keith MW, Anderson JR, Tyler J, Tyler DJ (2014) A neural interface provides long-term stable natural touch perception. Sci Transl Med 6(257):257ra138-257ra138

Tapu R, Mocanu B, Tapu E (2014) A survey on wearable devices used to assist the visual impaired user navigation in outdoor environments. In: 2014 11th international symposium on electronics and telecommunications (ISETC), pp 1–4. IEEE

Voss P, Lassonde M, Gougoux F, Fortin M, Guillemot JP, Lepore F (2004) Early-and late-onset blind individuals show supra-normal auditory abilities in far-space. Curr Biol 14(19):1734–1738

Voss P, Thomas ME, Cisneros-Franco JM, de Villers-Sidani É (2017) Dynamic brains and the changing rules of neuroplasticity: implications for learning and recovery. Front Psychol 8:274878

Wang G, Zhang X, Zhang J, Gruver WA (1995) Gripping force sensory feedback for a myoelectrically controlled forearm prosthesis. In: 1995 IEEE international conference on systems, man and cybernetics. Intelligent Systems for the 21st Century, vol 1, pp 501–504. IEEE

Chapter 2
Users's Requirements: Towards a User-Centered Approach

Abstract This chapter serves as the cornerstone of the book, focusing on the identification and analysis of user requirements critical for the design of wearable feedback systems and navigation aids. Through a user-centred approach, the chapter delves into the specific needs and preferences of visually impaired individuals as well as those with upper limb amputations. These insights are essential for developing effective and intuitive technological solutions. The findings presented here lay the groundwork for the subsequent development and implementation of wearable haptic feedback devices and indoor navigation systems, detailed in later chapters. By prioritizing user needs from the outset, this chapter aims to ensure that the resulting technologies are not only innovative but also practical and user-friendly, addressing the unique challenges faced by both visually impaired users and amputees.

2.1 User-Centred Approaches in Autonomous Navigation and Rehabilitation

As discussed in the previous chapter, sensory substitution technologies, particularly wearable haptic solutions, have significantly advanced over the past few decades for various applications. These technologies aim to replace lost sensory modalities with alternative sensory inputs, helping users perceive the environment in new ways. Despite these advancements, newly developed solutions often fail to meet user satisfaction and approval. Frequently, the devices are challenging to wear, bulky, and the stimuli they provide are not easily distinguishable or comprehensible to users, leading to their eventual abandonment.

One major issue is the limited involvement of users in the design phases of these devices, which often overlooks their needs and preferences. Involving users early and throughout the design process can lead to more intuitive, comfortable, and effective solutions. Therefore, a fundamental aspect of my work has been to place the user at the centre of the research. This involves studying the strategies they adopt in daily life and understanding how to integrate my devices to enhance and facilitate their everyday challenges. With this objective in mind, I aimed to reach as many users

F. Barontini, *Wearable Haptic Devices for Realistic Scenario Applications*,
Springer Series on Touch and Haptic Systems,
https://doi.org/10.1007/978-3-031-70539-7_2

as possible to identify the essential characteristics that my devices should have to meet their needs and requirements. By doing so, I hope to design wearable haptic devices that are not only technologically advanced but also practical and beneficial for real-world use.

2.1.1 Understanding Human Needs in Autonomous Navigation for Visually Impaired

Users' and accessibility requirements are fundamental to informing the design of the hardware and the algorithms for an effective indoor navigation system (Bach-y-Rita and Kercel 2003; Dakopoulos and Bourbakis 2009).

The first mandatory step for developing of satisfactory technological travel aids and intuitive User Interfaces (UIs) is to correctly understand how a blind person autonomously moves in an unknown environment. Choosing the appropriate approach is challenging, not only because of the special needs of VI people but also because of the considerable heterogeneity of visual impairment in the target group. Participatory design methods for VI users used scenarios, stimulating the dialogue between designers and blind users (Sahib et al. 2013). Another approach suggests workshops with blind and sighted people together (Okamoto 2009). Thereby, sighted participants were blindfolded to experience and subsequently discussed problems blind participants have to face.

The first step of our work is to collect information on how blind people usually move in unknown environments and which are their needs and requirements, we recruited and interviewed four visually impaired people. They were, respectively: a member of the Omero museum[1] (M, 60 years old—hereinafter referred to as S1, congenital blind), a member of "Lega del Filo d'Oro"[2] (F, 70 years old—hereinafter referred to as S2, blind since she was 12 years old), and two persons from the "Unione Italiana dei Ciechi ed Ipovedenti ONLUS (UIC)"—Pisa (Italy)[3] (M, 36 years old—hereinafter referred to as S3, congenital blind, and F, 42 years old—hereinafter referred to as S4, congenital blind). S1 and S3 use the white cane to walk in their everyday life while autonomously S2 uses a guide dog and sometimes a white cane, and S4 relies on an accompanying person in unknown environments. Furthermore, we also interviewed a teacher from "Lega del Filo d'Oro" specialized in training for the orientation and mobility of blind people.

[1] Omero museum is a museum specifically thought for visually impaired visitors, with dedicated and customized exhibition pathways, located in Ancona, Italy http://www.museoomero.it/.

[2] The Lega del Filo d'Oro is an Italian ngo, whose mission is to assist, teach and rehabilitate deaf-mutes http://www.legadelfilodoro.it/.

[3] Unione Italiana dei Ciechi ed Ipovedenti ONLUS (UIC)—Italian Association for the Blind—is a non-profit organization with legal entity governed by private law, which represents and protects the moral and material interests of the visually impaired people towards public administrations http://www.uici-pisa.it/uic-pisa/.

Table 2.1 Questions proposed during the preliminary interview

	Question
Q1	What do you commonly use as travel aid among guide dog, white cane and guide volunteer, and why?
Q2	If you walk through a corridor, where do you prefer to walk: in the center well away from possible obstacles close to the walls, or close to the walls?
Q3	If you are in a crowded place, hence with moving obstacles, which type of approach do you use?
Q4	Do you adopt different approaches if you are in an indoor or outdoor environment?
Q5	Which type of directional stimuli do you prefer to receive and where (body location)?
Q6	Do you prefer to experience on the arm a mild-constant cue along the duration of the movement and a more intense cue when there is a change of direction, or a stimulus only when needed?
Q7	In which way do you prefer to avoid obstacles?
Q8	Do you want to arrive as close as possible to the obstacle to understand which obstacle is, or do you want to stay far from it?
Q9	Using a travel aid system endowed with a camera, proximity sensors for obstacle identification and a device that conveys navigational cues in an unknown environment, which information would you like to receive as first and in an immediate way?
Q10	Do you prefer to have immediate indications about the presence of low or high obstacles?
Q11	Have you ever used a different travel aid with respect to the most common ones, for example vibrotactile-based devices?
Q12	Which type of stimulus do you prefer between auditory or tactile, and why?

All the questions are summarized in Table 2.1. The interviews were designed with the goal of identifying which features and functionalities a navigation system should ideally have and informing the design of the UI. For these reasons, questions focused on gathering information about: (1) mobility and orientation strategies used by blind people with no travel support or assistance (Questions from 1 to 4); (2) features that an indoor navigation support would be supposed to have (Questions from 5 to 10), and (3) the most suitable kind of stimulation modality to deliver the navigation cues to the end user, while she/he is walking, relying only on the ETA support (Questions 11 and 12).

The primary outcomes arising from the interviews are summarized in the following:

Mobility and Orientation Strategy

(1) *Paths*. Not smoothed but segmented routes are preferable for moving autonomously. Linear routes are usually preferred since they are more suitable for orientation.

(2) *Obstacles and mental map.* In unknown environments, e.g. hotel rooms, blind people are usually required to touch everything in an ordered manner to create an internal mental map of the location of the objects. On the other hand, while walking around, e.g. in a corridor, the main goal is to avoid obstacles on the ground; high-up obstacles, such as windows or shelves, should be avoided in a safely. In other terms, blind persons usually prefer to arrive very close to the object and touch it when they have to use it; otherwise, in case of an obstacle, they wish to avoid it.
(3) *Traditional white cane is necessary to support.* Participants S1, S2, S3 and the specialized teacher confirmed the importance of the traditional cane for a blind person, who usually moves around independently. The white cane helps detect obstacles, especially in potentially dangerous situations, like stairs or unexpected obstacles.

System Features

(4) *Wearable and hands-free system* While using a traditional white cane or a guide dog, the hands should be left free. Any communication should be provided through another part of the body. This confirms the observations in Dakopoulos and Bourbakis (2009), where the authors emphatize the characteristics of *Free-hands, Free-ears and Wearability* during the development of ETAs. The arm represents a good body location for haptic communication.
(5) *Reliable and able to detect any obstacle.* Many obstacles can be encountered when moving autonomously along a path or a route. The white cane can detect several obstacles, but unfortunately, not all of them. Obstacles located high-up and stairs may be particularly dangerous for a blind person. The system should be able to detect them and inform the users in a clear and unequivocal manner. The user should feel that the system is reliable for obstacle detection.

Feedback and Communication with the System

(6) *Haptic feedback.* Tactile feedback is preferred regarding auditory communication, as reported in Dakopoulos and Bourbakis (2009). Users motivated this choice with the need to have the auditory channel free while moving autonomously—especially outdoor—to perceive the surrounding environment. The audio feedback could be used for additional information requested on demand.
(7) *Simple and instantly recognizable instructions.* Simultaneous delivery of many types of information is not desirable. Clear and unequivocal instructions are needed for essential communications. Instantly recognizable indications should be used for crucial indications and important information types. For instance, a clear and single intense pressure to indicate a change of direction (instead of constant stimulation) or obstacle detection is preferable.

The requirements and needs expressed by the interviewed users were used to define the system specifications of the navigation system presented in Chap. 5.

2.2 Prosthetic User Able to Perceive: What Is Missing?

The development of user-centered health instruments, such as upper limb prostheses, must place the user at the center of the development cycle. Prosthetic limb users can provide significant insights to fulfill the primary goal of a prosthesis: to recreate or at least enhance mobility at home or in the community. A development process that does not incorporate user input and ignores the benefits of first-hand experience is likely to fail. Placing users at the heart of the design process ensures that the final product meets their actual needs and preferences, leading to higher acceptance and usability. This approach acknowledges that users are experts in their own experiences and can provide invaluable feedback that can guide the development of more effective and user-friendly prosthetic devices.

An established social scientific method to gather representative feedback in health research is the focus group (Schachter et al. 2011). A focus group consists of a limited number of expert participants engaging in an open discussion led by a moderator. By asking well-chosen open-ended questions, participants are encouraged to share information and involve other participants' opinions in their considerations. The moderator must skillfully guide the conversation to gather the necessary insights. The focus group discussion must be recorded and transcribed for further analysis. During the focus group sessions, participants were asked about their daily experiences with prosthetic devices, challenges they face, and their ideas for improvements. Questions such as *"What do you find most challenging about your current prosthesis?"* and *"What features would make a prosthesis more useful for you?"* helped to elicit detailed and meaningful responses.

During the European project SoftPro (https://softpro.eu), a survey was administered to individuals with upper limb loss at Mayo Clinic Hospital, Rochester, MN (USA). Its primary goal was to determine what type of feedback may be most useful and accepted by prosthesis users. The survey asked about tasks that might show improved performance with feedback, the most useful information to receive, and the best ways to provide that information. Demographic and prosthetic use information was also collected to provide a clear picture of the respondents.

The survey, partially presented in the Appendix, was administered as a pilot to a small pool of subjects ($n < 10$) by telephone. Participants were recruited based on the following inclusion criteria: male or female, amputation or dysmelia of at least one upper extremity, at least 16 years old, and a completed therapy phase post-amputation. Exclusion criteria included obvious cognitive deficits and a psychologically unstable state. Participants reported that a feedback system should provide information about the currently exerted grasping force, finger positions, and surface roughness. They also expressed a preference for receiving feedback when their fingers touch an object. Ideally, sensors should be integrated into the fingertips, with feedback delivered through the prosthesis shaft to the residual limb. The grasping force and reliability should be high, and the prosthesis should allow manipulation of both large and small objects, with the grasp adapting to increase contact area.

The insights gathered from the survey and focus groups were instrumental in shaping the design of the prosthetic devices. For instance, the feedback led to the development of prototypes that were lighter, more ergonomic, and equipped with advanced sensory feedback mechanisms that matched the users' preferences for touch and pressure feedback. This user-centered approach has broader implications for the field of prosthetics and rehabilitation. It demonstrates that involving end-users from the early stages of design can lead to the development of more effective and widely accepted solutions. This methodology can be applied to other areas of health technology, ensuring that new devices and systems meet the real-world needs of their users. By placing users at the forefront, this work highlights the importance of their involvement in the design process, leading to innovations that are not only technologically advanced but also practical and user-friendly.

Appendix

Following the survey on the feedback administered to the participants is presented. The questions were divided in three subcategories involving Demographics, Prosthesis Use, and Sensory Feedback. Only the Sensory Feedback is reported.

Sensory Feedback

(1) Do you have a sensory component to your prosthesis?
(2) [If the answer is no] Do you think your prosthesis would be improved if it had sensory feedback?
(3) Have you been involved in sensory feedback research?
(4) Which of the following activities do you think would be easier if you had sensory feedback in your prosthesis:

    a. Brush air
    b. Button shirt
    c. Tie shoes
    d. Use cup
    e. Use fork
    f. Write a word
    g. Use scissors
    h. Use phone
    i. Open a door knob

(5) Is there another activity not previously mentioned that you think would be easier with sensory feedback. (open ended question)
(6) There are several types of sensory feedback currently being investigated. The feedback gives the user information about location of the prosthesis in space, contact between the prosthesis and an object, and grip strength of the prosthesis.

Beside each description of modalities rank your interest in having this used in prosthesis:

a. Hand position
b. Object contact
c. Surface texture
d. Grasp force

(7) There are several ways to communicate the sensory information from the prosthesis to the user. Currently being investigated is a CUFF that goes on your stump (residual limb) and can stimulate the skin on that limb in a number of ways. Beside each description of stimulation please rank your interest in having this used in a prosthesis:

a. Vibration feedback
b. Skin stretch
c. Sliding stimuli on the limb
d. Band around the limb tightens or loosens

(8) Do you have any feedback for this survey? For instance were there other aspects of sensory feedback about which you want to comment?

References

Bach-y-Rita P, Kercel SW (2003) Sensory substitution and the human-machine interface. Trends Cognit Sci 7(12):541–546

Dakopoulos D, Bourbakis NG (2009) Wearable obstacle avoidance electronic travel aids for blind: a survey. IEEE Trans Syst Man Cybernet Part C (Appl Rev) 40(1):25–35

Okamoto M (2009) Possibility of participatory design. In: Human centered design: first international conference, HCD (2009) held as part of HCI international 2009, San Diego, CA, USA, July 19–24, 2009 Proceedings 1. Springer, Berlin, Heidelberg, pp 888–893

Sahib NG, Stockman T, Tombros A, Metatla O (2013) Participatory design with blind users: a scenario-based approach. In: Human-computer interaction-INTERACT 2013: 14th IFIP TC 13 international conference, Cape Town, South Africa, September 2–6, 2013, Proceedings, Part I 14, pp 685–701. Springer, Berlin, Heidelberg

Schachter D, Gilbert D, Wegner D, Hood BM (2011) Psychology, European edn

Chapter 3
Tactile Feedback in Prosthetics: A Pilot Study Exploring Different Haptic Modalities

Abstract This pilot study aims to determine which tactile sensory feedback, when combined with a soft robotic prosthetic hand, provides the most advantages for person with limb loss performing everyday tasks. Five volunteers with trans-radial amputations used the new SoftHand Pro (SHP) prosthetic hand to test a study protocol that evaluated single and combined feedback modalities while accomplishing various tasks. The tactile cues provided included contact information, grasping force, degree of hand opening, and combinations thereof. To transfer such feedback, different wearable systems were employed, utilizing either vibrotactile or force stimulation in a non-invasive, modality-matching approach. The results indicate a preference among amputees for single, rather than combined, feedback modalities. The choice of appropriate haptic feedback appears to be subject- and task-specific. Furthermore, in alignment with participants' feedback, force feedback with adequate granularity and clarity emerged as potentially the most valuable feedback type among those tested. Lastly, the study suggests that users prefer a prosthesis equipped with precise and clear feedback mechanisms.

3.1 Introduction

The human hand is a powerful tool to interact and operate with the environment. It allows to accomplish complex movements, from power to precision grasps and manipulation tasks. Moreover, it is an important communication vehicle for social interaction (Van Erp and Toet 2015). Losing a hand has devastating effects, especially concerning autonomy, limitation of capability, social life, and activities of daily living (ADLs). It changes people's life. In the last decades, the field of hand prosthetics developed rapidly with improvements in mechanics, dexterity, and control. The introduction of new myoelectrically driven hands obtained significant results for the recovery of the motoric functions actually lost with the amputation (Piazza et al. 2019). Despite this advancement, today's upper limb prosthesis users are still affected by relevant limitations that in the worst case may result in a rejection of the device. Among those, the lack of sensory feedback contributes to the limitation of an optimal control performance in everyday use (Culbertson et al. 2018). According to

F. Barontini, *Wearable Haptic Devices for Realistic Scenario Applications*, Springer Series on Touch and Haptic Systems,
https://doi.org/10.1007/978-3-031-70539-7_3

Biddis et al. (2007) and Cordella et al. (2016), the need for sensory feedback is one of the most important requirements of amputees with myoelectric and body-powered prostheses. In addition, amputees expressed also their dissatisfaction with the heavy dependence on visual feedback required to control current prostheses (Biddiss et al. 2007; Pylatiuk et al. 2007; Sensinger et al. 2020). Indeed, the restoration of haptic channels is an open challenge that covers several considerations. These are stimulation discomforts, the ability to relate the feedback to the grasping force and to proprioceptive information, technical constraints and pragmatic factors such as cost, weight and wearability.

Considerable efforts have been made in the last few years to find usable feedback solutions. In general, these solutions can be divided into two main categories: invasive and non-invasive feedback systems. Invasive systems interact directly with the amputee's nervous system and require surgical intervention. These systems could improve the prosthetic control and performance for amputees, however, such systems are mainly still at an experimental level since their translation into everyday life presents major challenges. For a complete review on invasive systems please refer to Kim (2022) and Bensmaia et al. (2020). In contrast, non-invasive solutions involve wearable haptic devices to convey various stimuli to the amputee. These systems rely on touch-mediated stimuli and can transfer information through residual limb squeeze, skin-stretch, vibration patterns and their combinations (see Dosen et al. Dosen (2021), Nemah et al. Nemah (2019), and Stephens-Fripp et al. Stephens-Fripp et al. (2018)). According to Schofield et al. (2014) three sub-categories of sensory feedback can be seen in hand prosthetics: they can be somatotopically matched, modality matched and based on a sensory substitution. Somatotopically matched feedback creates a natural signal comparable to the original sensation for the same body part. Modality matched feedback provides a stimulus similar to the original sensation, but at an alternative body site or extremity, e.g. mapping pressure at a finger on a similar pressure signal but to the forearm. Finally, a sensory substitution involves the transfer of information from a lost sense to the input channel of another sense. Usually, vibrotactile feedback systems fall into this category, since they transmit information, such as the grasping force, to the amputee's skin using frequency patterns as an alternative. Risi et al. (2019) investigated the use of vibrotactile feedback to improve limb motion control and showed induced changes in motion planning after training, but also associated with cognitive costs. Shah et al. (2019) investigated stimulation for body-machine interfaces to provide performance feedback to the user by conducting two psychophysical experiments to determine the effectiveness of vibrotactile perception via the arm. The results suggested that the arm could be a viable site to convey multivariate information via vibrotactile feedback for body-machine interfaces. In Clemente et al. (2015) the authors, exploiting the Discrete Event-driven Sensory feedback (DESC) policy, presented a device able to deliver short-lasting vibrotactile feedback to trans-radial amputees. The DESC-glove was tested by 5 trans-radial limb loss participants for one month at home. The results demonstrated the effectiveness of the device for prosthetic control in daily life conditions.

To the author's knowledge, only a few studies compared different feedback systems with amputees, to understand the advantages of stimuli during different ADLs. Patterson et al. (1992) compared vibratory and pressure feedback during a force matching task that involved grasping with a robotic arm. They analyzed the relative error using visual plus vibration feedback and visual plus pressure feedback. The results showed that the use of visual plus pressure feedback entailed in slightly lower relative error as regards the use of visual plus vibratory feedback. Similarly, Tejeiro et al. (2012) characterized the effects of vibration and pressure performance during object manipulation tasks. They demonstrated that participant performances increased with the addition of pressure or vibration stimuli compared to visual feedback alone. Bark et al. (2008) compared vibration and skin stretch stimuli performance during a virtual proprioception task. The results showed that skin stretch stimuli are easier to understand as proprioception information concerning the vibration stimuli.

A limit of the previous works was the fact that the limb loss users were not involved during the experimental task and the different sensor modalities always relied on visual feedback.

Based on these groundworks, the present study aims to evaluate and identify possible advantages of different non-invasive feedback for subjects with limb loss. We adopt the modality matching approach using two wearable devices, namely the Clenching Upper Limb Force Feedback (CUFF) and the VibroTactile. In the study, we test and investigate the usefulness and performance of different feedback modalities with five trans-radial users: contact information & roughness, grasping force, hand opening degree, grasping force combined with hand opening degree, and contact information combined with grasping force. More specifically, the amputees accomplished grasping tasks using upper limb prostheses in conjunction with wearable feedback devices without relying on visual feedback. At the best of the author knowledge this is one of the first studies that evaluates and collects insight from users who tested (and finally chooses) different haptic cues in coordination with the use of a myoelectric prosthetic hand. Although it presents a relatively small number of users it anyway provides, in the author's opinion, useful findings for future development of prosthetic feedback devices.

3.2 The Tactile Devices

The study aimed to deliver various types of stimuli (i.e., force, hand opening degree, first contact cue, and roughness) and their combinations (see Table 3.1) to subject with upper limb loss. To accomplish this, we employed the CUFF device and the Vibrotactile. Both of these devices can be seamlessly integrated with the soft robotic hand SoftHand Pro (SHP) (Catalano et al. 2012), enabling a comprehensive haptic experience for the users.

3.2.1 The Clenching Upper Limb Force Feedback Device

The Clenching Upper-Limb Force Feedback (CUFF) is a wearable device providing distributed mechano-tactile stimulation to the upper arm (Casini et al. 2015). Briefly, the CUFF device comprises a mainframe, which can be secured fastened to the user's arm with two Velcro straps, an actuation unit composed of two DC motors (DCX16S, Maxon, Switzerland AG, Alpnach Dorf, Swiss) with a gearbox ratio of $64 : 1$, and a belt that interfaces featuring with the skin of the human body. The device operates by utilizing motor rotation to provide two types of stimuli: squeeze and skin-stretch. When the motors rotate in opposite directions, the belt tightens or loosens, resulting in an adjustable pressure on the arm. On the other hand, when the motors rotate in the same direction (either medially or laterally), the belt slides, producing a skin stretch sensation that corresponds to the degree of hand opening. The users were instructed to associate the stimulus from the CUFF with the grip force and the hand opening degree of the SHP. To estimate the hand position, the readings of the two magnetic encoders (AS5045, ams-Osram AG, Premstaetten, Austria) were utilized. In addition, to estimate the grip force exerted by the hand, the residual current (RC_{meas}) was calculated. This parameter represents the difference between the estimated and the real current absorbed by the hand activating motor, as defined in Mura et al. (2018).

The process of mapping information from the robotic hand to the feedback device necessitates the establishment of a relationship between each hand value and the corresponding position value of the CUFF motor. To achieve this, a linear mapping strategy was selected to define the reference position (pos_{Cref}) of the CUFF motors. For the skin stretch stimulus, the encoder-detected motor position of the SHP $pos_{SHPmeas}$ was translated into the reference position. This translation adhered to two constraints: when $pos_{SHPmeas} = 0$ also the $pos_{Cref} = 0$ and when the $pos_{SHPmeas} = pos_{SHPmax}$, $pos_{Cref} = pos_{Cmax}$. Concerning the force stimulus, the translation of the residual current (RC_{meas}) into the reference position was conducted. Likewise, two constraints were identified: when $RC_{meas} = 0$ also the $pos_{Cref} = 0$, and when the $RC_{meas} = RC_{max}$ also the $pos_{Cref} = pos_{Cmax}$. Thus, the updated reference position of the CUFF motors can be expressed as follows:

$$pos_{Cref} = \left(\frac{pos_{SHPmeas}}{pos_{SHPmax}}\right) pos_{Cmax}, \tag{3.1}$$

$$pos_{Cref} = RC_{meas} * gain \tag{3.2}$$

respectively for the skin stretch stimulus (3.1), and for the force stimulus (3.2). The $gain$ was chosen heuristically equal to 0.4, while pos_{Cmax} was set at 800 ticks from the zero position as a result of testing that determined this position to be the maximum position well tolerated by a group of individuals.

3.2.2 The VibroTactile Feedback Device

The VibroTactile feedback system is a wearable digital device designed to relay real-time high-frequency contact information from a robotic hand to the user through vibrotactile stimuli (Fani et al. 2019). It consists of two main components: (i) the sensing unit and (ii) the actuation system. The sensing unit comprises two Inertial Measurement Units (IMUs) of type MPU-9250 (InvenSens, Inc, San Jose, USA). These IMUs are utilized to capture acceleration signals resulting from interactions with surfaces of grasping objects (Kuchenbecker 2010). For our specific application, we only utilized acceleration information within the range of $\pm 2g$. The actuation unit consists of two linear voice coils (NCC01-04-001-1X, H2W Technologies, CA, USA) selected for the system. Compared to the DC motors, voice coil actuators directly generate a linear movement, which can be easily transformed into a stimulation of the skin along the normal to the surface. To support the actuators and ensure proper alignment of the moving mass with the coil, a suitable frame was designed. An elastic fabric band secures the moving mass in place, preventing it from exiting the coil shaft while enabling efficient transmission of vibrations to the amputee. The actuator's coil axis is positioned perpendicular to the residual upper arm skin of the amputees.

The device is capable of conveying acceleration data recorded from the fingers of the SHP, which is related to contact and surface texture, to the user. To achieve this, the IMUs were placed on the side of the nails of the little and index fingers. This placement was chosen due to the fact that these fingers collectively cover a sufficient workspace for discriminating object size and that the index finger is commonly used to explore surfaces and textures.

To convert the acceleration measurements obtained from the IMUs into a Pulse Width Modulation (PWM) signal for controlling the voice coil, digital signal processing is employed. Initially, the accelerometer measurements are filtered to eliminate low-frequency (free-hand motion) and high-frequency noise, retaining only the frequency range that specifically activates human mechanoreceptors (i.e. the Fast Adapting Type I (FA-I), and the Pacinian corpuscles (Choi et al. 2012)). The frequencies sensed by these receptors ranged between 4 and 400 Hz, with a minimum threshold below 64 Hz for FA-I and between 128 and 400 Hz for Pacinian corpuscles (Johansson 1982). The vibrations generated by the SHP during opening and closing movements produce signals of interest within a threshold below 300 Hz. To accomplish this, a fourth-order Chebyshev Type I bandpass filter is employed, optimized for a bandwidth between 120 and 230 Hz, with signal reduction in the range of 140–210 Hz.

Furthermore, a dimensional reduction of the acceleration signal was performed. It is known that the human skin's response to vibration stimuli is independent of the acceleration direction. Therefore, we choose to stimulate the user's skin along a single direction orthogonal to the skin surface. Since acceleration is measured as a three-component vector, it needs to be reduced to a scalar value to drive the actuator. The technique used is the Sum of Components, computed as

$$a = |a_x| + |a_y| + |a_z| \tag{3.3}$$

where a is the output value, used to drive the coil, and a_x, a_y, a_z are the three components measured by the accelerometer. For more information regarding signal processing and dimensional reduction please refer to Fani et al. (2019).

3.2.3 *The SoftHand Pro*

The SoftHand Pro (Catalano et al. 2012) is the prosthetic version of the robotic Pisa/IIT SofHand and was used as the unique prosthetic hand for the present experiments. The Pisa/IIT SoftHand is a kinematic synergy inspired hand with 19 DOFs resulting from two ext/flex and one abd/add DOFs at the first phalange and from three ext/flex and one abd/add DOFs at each of the four remaining phalanges. All joints are manufactured as rolling contact joints with elastic ligaments ensuring an accurate closure of the hand when actuated and allowing for unharmful interactions with humans. Elastic ligaments allow for the deformation of phalanges and ensure their return to the initial position. A single cable runs through all joints to simultaneously flex and abduct the fingers upon actuation. The hand is actuated by a single 15 W DC motor (DCX22S, Maxon, Switzerland AG, Alpnach Dorf, Swiss) driving the cable to move the fingers on the path of the first synergy as described by Santello et al. (1998). The SHP can easily adapt to the environment as all of the fingers conform to the shape of grasped objects which augments its grasping capabilities. In addition, the SHP interfaces with commercially available surface electromyography (EMG) sensors ($13E200 = 50$ MyoBock, Otto Bock, Duderstadt, Germany). These sensors detect the action potentials of corresponding muscle fiber membranes during muscle contraction. They were placed on the flexors and extensors of the forearm to activate the closing and opening of the SHP, respectively.

3.3 Feedback Modalities Testing and Analysis

As mentonioned previously, in this work we want to compare different feedback modalities and comparing their effectiveness with prosthesis users, and for this reason, we considered the devices individually and their combination. More specifically we chose:

- force feedback only, using the CUFF in force modality (hereinafter referred to as CF)
- proprioception feedback only, using the CUFF in proprioception modality (hereinafter referred to as CP)
- vibration stimuli, using the Vibrotactile (hereinafter referred to as V)

Table 3.1 Delivered information to the user, the stimulus provided to the arm and used haptic device. The haptic devices are referred to as follows: CF for the CUFF in Force modality, CP for the CUFF in Proprioception modality, and V for VibroTactile. Consequently, CF + CP indicates the combination of the CUFF in Force modality with the CUFF in Proprioception modality, and CF + V indicates the combination of the CUFF in Force modality with the VibroTactile. These terminology will be used also in the text

Information to deliver	Stimulus	Haptic feedback		Acronym
Grasping force	Squeeze	CUFF in force modality		CF
Hand opening degree	Skin-stretch	CUFF in proprioception modality		CP
Contact information and roughness	Vibration	VibroTactile		V
Grasping force + Hand opening degree	Squeeze + Skin stretch	CUFF in force modality + CUFF in proprioception modality		CF+CP
Grasping force + Contact information and roughness	Squeeze + Vibration	CUFF in force modality + VibroTactile		CF+V

- a combination of force and vibration feedback, using the CUFF in force plus the vibrotactile (hereinafter referred to as $CF + V$)
- a combination of force and proprioception feedback, using the CUFF in force plus proprioception modality (hereinafter referred to as $CF + CP$)

as shows in Table 3.1.

3.3.1 Research Participants

Five trans-radial amputees S1 - S5 gave their informed consent and were initially included (mean age ± SD: 63.2 ± 11.7, 1 female). These were four unilateral and one bilateral amputee. Exclusion criteria were cognitive and neurological deficits affecting the experiments. The procedures described in this paper were approved by the Local Ethics Committee of the Hannover Medical School (7444/18). The experiments were divided into two sessions: during the first session, the participants with limb loss should familiarize themselves with the SHP and test all feedback system modalities presented in Table 3.1. At the end of the session, they should select the feedback modalities they subjectively felt most comfortable with. Later, they should test these chosen system modalities in a second session.

All amputees participated in the first session, but due to the COVID-19 pandemic only two of them (mean age ± SD: 53.5 ± 6.4, 1 female) could be reincluded into the second session.

3.3.2 Experimental Tests and Questionnaires

To evaluate the performance of the participants we identify specific tasks and questionnaires.

The participants were instructed to perform upper-limb dexterity test, i.e. the Activities Measure for Upper Limb Amputee (AM-ULA) (Resnik et al. 2013), and the Jebsen Taylor Hand Function Test (Hackel et al. 1992) using their own prosthesis. The AM-ULA test comprises 18 tasks that assess various aspects such as task completion, speed, movement quality, skill in prosthetic use, and level of independence. This measure has demonstrated excellent internal consistency, good inter-rater reliability, test-retest reliability, and showed known-group and convergent validity. Additionally, the Jebsen Taylor Hand Function Test (JTHFT) evaluates ADL performance based on the completion time of 7 simulated tasks, including writing, feeding, and manipulating both large and small objects. Redundant tasks in both tests were performed only once.

To test the feedback modalities, we design three manipulation tasks:

1. Recognition Object task: the participants had to recognize and grasp five objects named by the therapist, without visual feedback. The objects chosen from Table 3.2, exhibited variation in sizes, shapes, textures and softness. All the objects in Table 3.2 were selected as they are commonly used in activities of daily living.
2. Size Discrimination task: the participants had to sort five wooden spheres (diameter of 20, 40, 50, 60 and 80 mm) from the smallest to the largest.
3. Softness Discrimination task: the participants had to sort five balls (i.e. tennis, massage, gym balls) from the softest to the stiffest.

For all tasks, the objects were positioned inside a box ($W \times H \times D = 0.7 \times 0.5 \times 0.3$ m; see Fig. 3.2). The box was enclosed by a curtain on the subject's side and left open on the therapist's side. The curtain concealed the objects from the participants' view, enabling the experimenter to observe, record videos, and evaluate the exploration process. Throughout the task execution, the subjects engaged in conversation with the therapist, introducing an intentional distraction factor. During the tests, the subjects were comfortably seated at an adjustable height table, with the closed box placed in front of them. The subject wore the CUFF and the VibroTactile device on the upper arm, as shown in Fig. 3.1. It is worth noticing that the CUFF device was calibrated around each user's arm before starting the task. The calibration of the CUFF involved an auto-adjustment of the fabric belt around the participantâŁ™s arm in which the fabric was tightened until the two motors reached the stall condition and then released leaving the absorbed current close to zero. This process guarantees the complete contact of the belt on the arm. Thanks to the design of the frame housing the

Table 3.2 The objects used for recognizing object task

	Low	Medium	High
Stiffness	• Thin scarf • Stuffed animal • Wool ball	• Full pack of tissues • 3/4 full 1.5 l water bottle • Carton box (10 × 10 × 10 cm)	• Pineapple • Hard cover book (10 × 20 × 2 cm) • Screwdriver • Fork
Size	• Full pack of tissues • Thin scarf • Screwdriver • Fork	• Carton box (10 × 10 × 10 cm) • Wool ball • Hard cover book (10 × 20 × 2 cm)	• Pineapple • 3/4 full 1.5 l water bottle • Stuffed animal
Roughness	• Full pack of tissues • Thin scarf • Wool ball	• Pineapple • Hard cover book (10 × 20 × 2 cm) • Stuffed animal • Carton box (10 × 10 × 10 cm)	• 3/4 full 1.5 l water bottle • Screwdriver • Fork

actuator, no calibration phase was required for the VibroTactile device. Finally, the subjects accomplished two surveys: a qualitative questionnaire evaluated on a 7-point Likert scale (Prattichizzo et al. 2013), and a survey focused on the chosen feedback modality. The qualitative questionnaire consists of a series of inquiries related to the system and the experimental task. Participants were asked to assign scores ranging from 1 (totally disagree) to 7 (totally agree). This approach represents a common procedure to evaluate devices for assistive robotics and Human-Robot Interaction. The questionnaire was structured as follows:

- Questions Q1–Q11 focus primarily on the analysis of the different feedback modalities tested by the subjects. This includes assessments of both single modality and combined modality.
- Questions Q12–Q16 are centred around evaluating the subjects' performance with the feedback.
- Questions Q17–Q19 explore potential applications of the various feedback modalities in everyday scenarios.

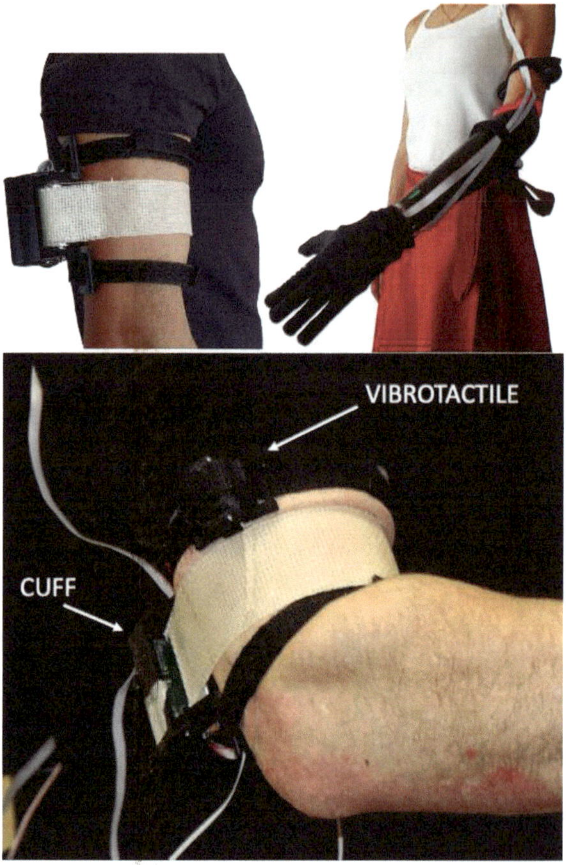

Fig. 3.1 The CUFF, the VibroTactile devices and their combinations applied on the user's arm, during this investigation

- Question Q20 investigates the perception of the current version of the feedback.
- Question Q21 pertains to the mental workload required to comprehend the stimuli conveyed by the devices.

The survey regarding the feedback was created ad hoc for the experimental session. The participants were required to answer the following questions:

1. Why did you choose this feedback device/modality?
2. In which everyday situation this feedback device could help you?
3. Which of the feedback devices presented to you would be least helpful, and why?
4. What changes would be necessary to make the devices you did not choose more useful to you?

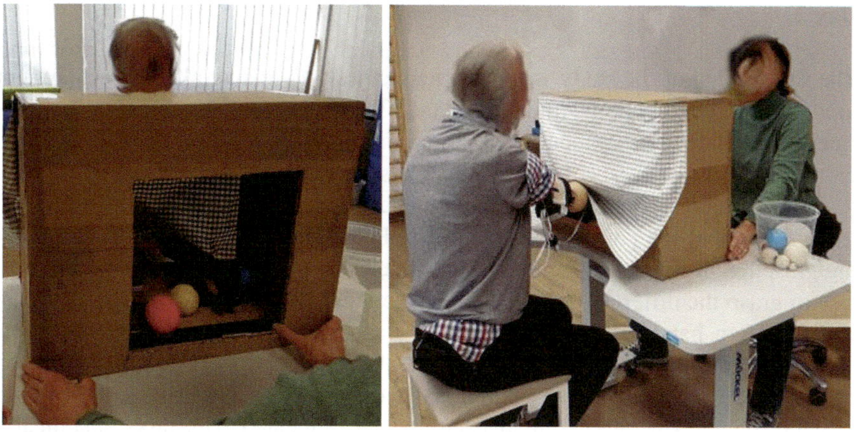

Fig. 3.2 Experimental setup: on the left the investigator view, on the right and view direction of the amputee

3.3.3 *Experimental Protocol of the First Session*

The first session was divided into three days.

Day 1

The first day of the study was dedicated to conducting dexterity tests and familiarizing the participants with the control and usage of the SoftHand Pro. In the first phase, the participants were instructed to perform an upper-limb dexterity test, i.e., the Activities Measure for Upper Limb Amputee (AM-ULA) (Resnik et al. 2013), with their own prosthesis. After this test, the SoftHand Pro was configured and attached to the user's socket. Under the supervision of a specialized physical therapist trained in prosthetic training for upper limb amputees, the participants underwent a 30-minute familiarization period. During this phase, they learned how to control and utilize the new robotic hand. The familiarization process included picking up and stacking objects including blocks and collapsible drinking cups, as well as picking up and placing down objects of different sizes, shapes, and materials. After this period, the participants repeated the AM-ULA using the SHP. The execution time was recorded for each test and each robotic hand used. In addition to evaluating dexterity, the initial day of the study also aimed to compare the agility in utilizing the SoftHand Pro relative to participants' own prostheses, as this aspect could potentially impact the outcomes of the comprehensive experiment involving the feedback devices.

Day 2

The second day of the study was dedicated entirely to testing all feedback system modalities in conjunction with the SHP (see Table 3.1). A 60-minute slot was allocated per each feedback modality, and a minimum break of 30 min was given to the participants between tests of two feedback system modalities.

Each slot consists of a training session and an experimental session. During the training session, the participant used the SHP with the feedback for 30 min playing/grasping all the objects presented in Table 3.2 with visual feedback. They were free to grasp the different objects and test the feedback device in various everyday situations (e.g. handshake, stacking objects, grasping collapsible drinking cups). Following the training session with the visual feedback, a 10-minute training with a black box was conducted. The subject explored five objects chosen from Table 3.2 without visual feedback. The final 20 min of each slot were allocated to the tasks, specifically the Recognizing Objects, Size Discrimination, and Softness Discrimination tasks.

Throughout the day, participants were closely accompanied by the therapist, who managed the training sessions and supervised the test sessions. The order of the feedback combinations was randomized across subjects. During the experiment we record the success, defined as the number of attempts required to retrieve the correct object, and the time of success, defined as time the participants believed that they had solved the task. At the end of each feedback modality, the subjects accomplished the qualitative survey. Corresponding questions can be taken from Table 3.5. At the end of the day, participants were provided with an additional questionnaire to compare the different systems and choose their preferred feedback modality, stating their reasons.

Day 3

On the last day, the feedback system chosen by the participant during the second day was tested more in-depth. The subjects carried out the AM-ULA and the Jebsen test with the SHP integrated with the feedback system chosen during Day 2. Subsequently, a second repetition of the two tests without feedback, aimed to assess the potential presence of learning effects during the test execution.

3.3.4 Experimental Protocol of the Second Session

At the end of the first experimental session, two feedback systems resulted as the preferred ones: the CUFF in force modality and the VibroTactile. The two feedback system modalities chosen during the first session were retested in this session lasting one day. Therefore, the Recognizing Object task was accomplished again with a separate use of feedback modalities. The qualitative questionnaire was finally surveyed again (Table 3.5). The experimental setup and procedure were the same as the first session.

3.4 Results

3.4.1 *First Experimental Session*

From the first session of the experiment, as previously anticipated, we observed a preference for the two feedback modalities, i.e. the CUFF in force modality and the VibroTactile. Table 3.3 presents the results of the AM-ULA test conducted in three modalities: (1) with the prosthesis utilized by users in their daily lives, (2) with the SHP, and (3) with the SHP plus the haptic feedback.

Table 3.4 displays the outcomes of the Recognizing Object task. It is worth noting that it has been decided to present only the result obtained during this task. This choice is grounded in the author's conviction that this particular task was the most informative in demonstrating how each feedback modality could be utilized to provide distinct information.

Table 3.5 provides a comprehensive overview of the qualitative questionnaire. Additionally, Table 3.7 presents the results of the Jebsen test, comparing performance when using the SHP both with and without the feedback modality chosen on Day 2.

Lastly, Table 3.8 succinctly summarizes the impressions and comments expressed by the subjects at the end of Day 2, including an investigation into their motivations for selecting specific feedback modalities.

Table 3.3 The table show the results of the AM-ULA test accomplished in three modalities: with their own prsthesis, with the SHP and with the SHP + Feedback

Subject	Own prosthesis	SHP	SHP+ Feedback
S1	11.11	14.44	13.89
S2	10.83	11.39	11.11
S4	13.33	15.00	15.00
S5	16.39	16.11	17.50

Table 3.4 The table shows the results of the feedback task: # non completed, - non performed, S means Subject, CF means CUFF in Force modality , CP means CUFF in Proprioception modality, V VibroTactile, CF+V CUFF in Force plus the Vibrotactile, CF+CP CUFF in Force plus Cuff in Proprioception

	Recognizing object task									
	Execution time [s]					N of errors				
S	CF	CP	V	CF+V	CF+CP	CF	CP	V	CF+V	CF+CP
S1	37	60	63	168	–	1	0	0	0	–
S2	257	154	88	110	300	2	0	0	0	3
S3	128	232	331	153	–	1	0	0	3	–
S4	132	105	126	90	152	2	0	2	0	3
S5	104	106	159	111	–	2	2	2	2	–

Table 3.5 Results of the qualitative questionnaires evaluated on a 7-point Likert scale (1: Strongly disagree, 7: Strongly agree). First session values are means and SDs resulting from amputees S1 - S5. Second session results are individual score values of S4 and S5 (grasping force and contact information feedback convey using the CUFF in force modality and the VibroTactile). Unperformed feedback system modalities are indicated with dots

	Questions	First session		Second session	
		Mean	Std. Dev.	A4	A5
Q1	[V] The vibrations provided by the actuators were not confusing	5.0	0.9	7	5
Q2	[V] The vibrations provided by the actuators were not disturbing	6.8	0.4	7	6
Q3	[V] The vibrotactile device allowed different level of roughness to be distinguished	5.2	1.3	3	4
Q4	[V] The vibrotactile device allowed contact with the external objects to be clearly detected	4.0	1.9	5	4
Q5	[CF] I was able to distinguish different levels of force through the cutaneous device	3.0	2.1	6	4
Q6	[CF] The maximum force elicited pain	2.8	2.2	1	1
Q7	[CP] The belt generated pain when used for a long time	2.8	1.9	–	–
Q8	[CP] The proprioception feedback (sliding movement) was clear	4.6	1.4	–	–
Q9	[CP] I was able to understand the position of the hand, also after a distraction (perception retention)	5.0	1.4	–	–
Q10	[CF+V] The two stimuli (vibration and force) were clearly distinguishable	4.6	1.4	–	–
Q11	[CF+V] The stimuli together were confusing	3.6	1.6	–	–
Q12	[CF+CP] The two stimuli (force and proprioception) were clearly distinguishable	4.5	0.5	–	–
Q13	[CF+CP] The stimuli together were confusing	2.0	0.0	–	–
Q14	I felt I performed better while receiving feedback by the cutaneous device	3.8	2.7	2	4

(continued)

Table 3.5 (continued)

	Questions	First session		Second session	
		Mean	Std. Dev.	A4	A5
Q15	I felt hampered by the cutaneous device	3.0	2.3	6	5
Q16	The stimulus provided by the cutaneous device was strange/weird	3.0	2.3	2	4
Q17	It was easy to perform the tasks while wearing the cutaneous device	3.6	1.9	6	4
Q18	The noise generated by the actuators interfered with the haptic perception	4.2	2.6	1	4
Q19	The feedback given was useful during the everyday life	4.6	2.0	4	3
Q20	The feedback given was useful during work	4.6	2.0	6	2
Q21	The feedback given was useful for interaction with other people	4.4	2.1	2	5
Q22	The design of haptic feedback device must be changed	6.0	1.5	7	7
Q23	I had to pay attention to the cutaneous device instead of the task to feel the stimuli	2.6	2.2	1	2

It is worth noticing that participant A3 has been excluded from the AM-ULA as well as from the Jebsen test, due to personal circumstances that prevented the completion of the three-day experimental session. However, we thought interesting to present the results obtained with the utilization of feedback in the Recognizing Object task. Furthermore, given the limited number of participants in the study, it was not feasible to conduct any form of statistical analysis.

3.4.2 Second Experimental Session

The second session of the experiment has been performed so far by only two participants with limb loss. For this reason, it was not possible to accomplish a statistical analysis. The VibroTactile and the CUFF in force modality were the feedback chosen from the first session of experiments. Table 3.6 shows the results for the recognizing object task. At the end of the experiments, both participants underwent the qualitative questionnaire again (see *Second Session* in Table 3.5).

Table 3.6 The table show the results of the Jebsen test in seconds

Task	SHP Post training				SHP + Feedback			
	S1	S2	S4	S5	S1	S2	S4	S5
Simulated page turning	35,4	22,62	28,33	34,15	31,07	26,53	27,03	36,87
Lifting small object	82	124	61,7	124,77	59,5	77	58,2	62,12
Simulated feeding	98	67	25,82	34,69	30,57	109	22,4	31,94
Stacking checkers	47	297	69,7	84,82	92	179	74,6	#
Lifting light can	13,46	16,9	12,6	16,11	19,23	16,11	13	24,05
Lifting heavy can	15,3	16,3	10,5	13,35	13,94	13,8	14,6	17,69

Table 3.7 The table shows the results of the recognizing object task of the second session of experiment accomplished by two subjects

Recognizing object task				
	Execution time [s]		N of errors	
Subject	CF	V	CF	V
S4	96	#	0	#
S5	17	103	0	1

Table 3.8 The table shows the chosen feedback for all the participants. The motivations for the choices were also presented

Subject	Chosen feedback	Motivation
S1	VibroTactile	Impressed to get a feedback when touching an object. The VibroTactile provided him with a clear information/feedback. He would have chosen the force feedback, if the signal/feedback had been clearer and the range larger
S2	CUFF in proprioception modality	The clearest and most helpful information for the use of the prosthetic hand
S3	VibroTactile	He would have chosen the force feedback, if the signal/feedback had been clearer and the range larger
S4	CUFF in force modality + VibroTactile	The most potential has the combination force + VibroTactile, but he would need time to practice and to understand the feedback information
S5	VibroTactile	The CUFF was disturbing

3.5 Discussion

3.5.1 *Outcomes of the First Experimental Session*

Table 3.3 presents the results of the AM-ULA test. It is noteworthy that the task execution times exhibit minimal variation between the hand commonly used by individuals in their daily lives and the SHP. This observation suggests that even with a limited amount of training, the participants were able to effectively control and utilize the SHP. However, regarding the performance of the AM-ULA test in the SHP + feedback modality, the task completion times are not significantly different from those observed when using the SHP alone. Consequently, it can be inferred that the feedback neither worsens nor improves the overall performance.

Regarding the Recognizing Object Task, due to the limited number of subjects, it was not feasible to conduct a comprehensive statistical analysis. However, from a qualitative standpoint, it can be noted that all subjects successfully completed the task when utilizing information from a single feedback modality. About the combined modalities, no significant difficulties were encountered in combining force feedback with vibration. However, it is worth mentioning that only two subjects managed to accomplish the task in the CF+CP mode. The remaining subjects expressed an inability to discern between the two stimuli and subsequently declined to proceed with the task.

In general, the prevailing approach adopted by the subjects involved in identifying object presence based on the feedback received during the initial contact. Once the object's position was determined, the subjects proceeded with exploration by means of grasping and manipulating the object. This strategy proved to be successful, as confirmed by the qualitative questionnaire, especially when the vibrotactile device was worn. The utilization of vibrotactile feedback for object localization was further substantiated by the responses to Question 4 (see Table 3.5), wherein it became evident that the provided vibrations facilitated effective contact with external objects. Less strong is the result obtained with the CUFF in force modality. What emerged from questions Q4 and Q5 is that the stimulus provided was not enough strong to allow to distinguish the different levels of force. The CUFF in proprioception modality (question from Q6 to Q7) revealed that the stimulus was comfortable and easy to understand, but in general, the subjects preferred to receive information about the force grasp instead of the position of the hand. Regarding the combination of the feedback modalities, comments are more neutral without showing any good or bad opinions by the participants. More specifically questions Q8 and Q9 referred to the combination of force with the VibroTactile feedback. The subjects were able to distinguish the two stimuli with a mean score of 4.6 ± 1.4 SD, but at the same time, the combination of the two stimuli was a little bit confusing with a mean score of 3.6 ± 1.6 SD. The same conclusions could be deducted from the combination of force and proprioception modality: the stimuli were not confusing but neither easy to discriminate (4.5 ± 0.5). These outcomes were in line with the results obtained in Kim and Colgate (2012) where the simultaneous display of two haptic channels did not enhance the control of the prosthesis. The participants did not feel to performed better with the working feedback device (questions Q12 3.8 ± 2.7, and Q15 3.6 ± 1.9) even if these results are in contrast with the objective results related to the time needed for task execution and the success rate (see Table 3.4). Moreover, the usefulness of the feedback during work, interaction with people and daily tasks received a positive score, but the design of the feedback needs to be improved. These outcomes may be explained considering the task and the experimental protocol designed. A test with longer training and an extended experimental session may be improved to understand how to use the feedback in the everyday life.

All the subjects, out of one, found the feedback potentially useful in everyday life both during work and in the interaction with others. The same four subjects stated that the feedback could be perceived even without focusing on the device output, maintaining the attention on the task. Looking at a detached comparison between devices, it can be observed that there is no specific trend, but it changes greatly between subjects, suggesting that the quality of the perception of the provided feedback is very subject-dependent.

Other important information coming out from the first session are the qualitative comments of subjects. The participants agreed on the fact that the usage of the CUFF in force modality could be useful for daily-living task execution, but the main comment was related to the provided stimulus amplitude, which resulted to be not wide enough to give an appropriate granularity of the feedback. The force feedback system was unable to effectively communicate different levels of force due to the inadequate mapping of residual current values. The chosen linear mapping and the two constraints resulted in a on-off activation pattern for the belt. Consequently, participants perceived either a strong grip, regardless of the object's rigidity or softness, or the release of the belt when the hand was opening. Based on these findings, one of the future developments we have decided to pursue is the implementation of a logarithmic mapping to enable participants to distinguish between various force levels. The CUFF in proprioception modality resulted useful mainly to understand if the hand was moving or not, for the proprioception feedback on how far the hand was opened. Among all the tasks proposed in the Jebsen test, the Small Objects Lifting task is the one in which the use of the feedback strongly improved the results with all the feedback systems. On the other side, the worst task in terms of improvement given by the feedback device is the Stacking Checkers task, where an increase in the execution time can be noted for all amputees except one.

3.6 Limitation and Future Works

This work reports an investigation of different feedback modalities to find out which is the most effective solution. Despite we can deduce some useful information regarding the user requirements and needs, we are aware that our approach presents some limitations, which we would like to address in future work. The main aspects we would like to further develop are: (1) the cohort size, (2) the hardware architecture.

Regarding the first aspect, leading an extensive campaign with limb loss participants, supported by statistical analysis, could increase the information about feedback preferences and types of stimuli required. A more in-depth analysis of the CUFF device and its mapping strategy (aspect 2), could improve the reduced capacity of the force feedback to convey squeezing stimuli. The adoption of a non-linear mapping, like a logarithmic one, could be a better choice since it can be used to obtain an amplifying effect over the mapped domain, and convey to the user a more effective squeezing sensation. Moreover, the miniaturization of the device with target integration inside the prosthesis could improve the control of the prostheses during work, daily living activities, and social interaction. Future works could include the usage of feedback with a different type of robotic hand. Furthermore, we would also like to conduct a study on the potential delay introduced in the system by the devices, comparing it to the existing delay of the EMG system.

This could suggest that the new feedback modalities are effective in delivering information without the use of visual feedback but probably, longer training is required to familiarize with the tactile stimulation. Future investigations could include performance comparisons between the use of visual feedback plus a vibration or force feedback concerning the visual feedback only.

3.7 Conclusions

In this work, we presented a comparison between the use of vibration, skin stretch and multi modal stimuli. The target was to understand which feedback modality was preferred by limb loss participants for better control of the prosthetic hand without the use of visual feedback. Notwithstanding the small number of included amputees for the collection of further information, the results of the experiments showed that the single modality feedback is preferred to the combination of multiple feedback. The combination of different modalities do not improve the performance during the task execution and the stimuli were difficult to understand. On the other hand, the participants found the grasping force and the roughness feedback helpful to accomplish the task of daily activities, suggesting an easy integration by the user of the device. As said previously, the choice of feedback with respect to the other was very subject- and task- dependent. Finally, the qualitative evaluation underlines that prosthetic solutions could be improved where the subject can choose the type of feedback to integrate into his prosthesis.

Acknowledgements The authors would like to thank the volunteers for their efforts and patience during the study; Wiebke von Kloth from John & Bamberg for physiotherapeutically trainings with the subjects; Benjamin Fleischer-Lück, Juliana Clavijo Rincón, and Carina Gempfer for assistance in data collection and processing; Cristiano Petrocelli and Mattia Poggiani for the hardware and software assistance.

References

Bark K, Wheeler JW, Premakumar S, Cutkosky MR (2008) Comparison of skin stretch and vibrotactile stimulation for feedback of proprioceptive information. In: Symposium on haptic interfaces for virtual environment and teleoperator systems. IEEE, pp 71–78

Bensmaia SJ, Tyler DJ, Micera S (2020) Restoration of sensory information via bionic hands. Nat Biomed Eng 1–13

Biddiss E, Beaton D, Chau T (2007) Consumer design priorities for upper limb prosthetics. In: Disability and rehabilitation: assistive technology, vol 2, no 6, pp 346–357

Casini S, Morvidoni M, Bianchi M, Catalano M, Grioli G, Bicchi A (2015) Design and realization of the cuff-clenching upper-limb force feedback wearable device for distributed mechano-tactile stimulation of normal and tangential skin forces. In: IEEE/RSJ international conference on intelligent robots and systems (IROS). IEEE, pp 1186–1193

Catalano MG, Grioli G, Serio A, Farnioli E, Piazza C, Bicchi A (2012) Adaptive synergies for a humanoid robot hand. In: 12th IEEE-RAS international conference on humanoid robots (Humanoids 2012). IEEE, pp 7–14

Chinello F, Pacchierotti C, Tsagarakis NG, Prattichizzo D (2016) Design of a wearable skin stretch cutaneous device for the upper limb. In: IEEE haptics symposium (HAPTICS). IEEE, pp 14–20

Choi S, Kuchenbecker KJ (2012) Vibrotactile display: perception, technology, and applications. In: Proceedings of the IEEE, vol 101, no 9, pp 2093–2104

Clemente F, D'Alonzo M, Controzzi M, Edin BB, Cipriani C (2015) Non-invasive, temporally discrete feedback of object contact and release improves grasp control of closed-loop myoelectric transradial prostheses. IEEE Trans Neural Syst Rehabilitation Eng 24(12):1314–1322

Cordella F, Ciancio AL, Sacchetti R, Davalli A, Cutti AG, Guglielmelli E, Zollo L (2016) Literature review on needs of upper limb prosthesis users. Front Neurosci 10(209)

Culbertson H, Schorr SB, Okamura AM (2018) Haptics: the present and future of artificial touch sensation. Annu Rev Control, Robot, Autonomous Syst 1:385–409

Dosen S, Prahm C, Amsüss S, Vujaklija I, Farina D (2021) Prosthetic feedback systems. In: Bionic limb reconstruction. Spring Cham, pp 147–167

Fani S, Di Blasio K, Bianchi M, Catalano MG, Grioli G, Bicchi A (2019) Relaying the high-frequency contents of tactile feedback to robotic prosthesis users: design, filtering, implementation, and validation. IEEE Robot Autom Lett 4(2):926–933

Hackel ME, Wolfe GA, Bang SM, Canfield JS (1992) Changes in hand function in the aging adult as determined by the Jebsen test of hand function. Phys Therapy 72(5):373–377

Johansson RS, et al (1982) Responses of mechanoreceptive afferent units in the glabrous skin of the human hand to sinusoidal skin displacements. Brain Res 244(1):17–25

Kim K (2022) A review of haptic feedback through peripheral nerve stimulation for upper extremity prosthetics. Curr Opin Biomed Eng 100368

Kim K, Colgate JE (2012) Haptic feedback enhances grip force control of sEMG-controlled prosthetic hands in targeted reinnervation amputees. IEEE Trans Neural Syst Rehabilitation Eng 20(6):798–805

Kuchenbecker Katherine J et al (2010) VerroTouch: high-frequency acceleration feedback for telerobotic surgery. In: International conference on human haptic sensing and touch enabled computer applications. Springer, Berlin, Heidelberg, pp 189–196

Mura D, Barbarossa M, Dinuzzi G, Grioli G, Caiti A, Catalano MG (2018) A soft modular end effector for underwater manipulation: a gentle, adaptable grasp for the ocean depths. IEEE Robot Autom Mag 25(4):45–56

Nemah MN, Low CY, Aldulaymi OH, Ong P, Qasim AA (2019) A review of non-invasive haptic feedback stimulation techniques for upper extremity prostheses. Int J Integr Eng 11(1)

Patterson PE, Katz JA (1992) Design and evaluation of a sensory feedback system that provides grasping pressure in a myoelectric hand. J Rehabil Res Dev 29(1):1–8

Piazza C, Grioli G, Catalano MG, Bicchi A (2019) A century of robotic hands. Annu Rev Control, Robot, Autonomous Syst 2:1–32

Prattichizzo D, Chinello F, Pacchierotti C, Malvezzi M (2013) Towards wearability in fingertip haptics: a 3-dof wearable device for cutaneous force feedback. IEEE Trans Haptics 6(4):506–516
Pylatiuk C, Schulz S, D ö derlein L (2007) Results of an internet survey of myoelectric prosthetic hand users. Prosthetics Orthotics Int 31(4):362–370
Resnik L, Adams L, Borgia M, Delikat J, Disla R, Ebner C, Walters LS (2013) Development and evaluation of the activities measure for upper limb amputees. Arch Phys Med Rehabilitation 94(3):488–494
Risi N, Shah V, Mrotek LA, Casadio M, Scheidt RA (2019) Supplemental vibrotactile feedback of real-time limb position enhances precision of goal-directed reaching. J Neurophysiol 122(1):22–38
Santello M, Flanders M, Soechting JF (1998) Postural hand synergies for tool use. J Neurosci 18(23):10105–10115
Schofield JS, Evans KR, Carey JP, Hebert JS (2014) Applications of sensory feedback in motorized upper extremity prosthesis: a review. Expert Rev Medical Dev 11(5):499–511
Sensinger JW, Dosen S (2020) A review of sensory feedback in upper-limb prostheses from the perspective of human motor control. Front Meurosci 14(345)
Shah VA, Casadio M, Scheidt RA, Mrotek LA (2019) Spatial and temporal influences on discrimination of vibrotactile stimuli on the arm. Exp Brain Res 237(8):2075–2086
Stephens-Fripp B, Alici G, Mutlu R (2018) A review of non-invasive sensory feedback methods for transradial prosthetic hands. IEEE Access 6:6878–6899
Tejeiro C, Stepp CE, Malhotra M, Rombokas E, Matsuoka Y (2012) Comparison of remote pressure and vibrotactile feedback for prosthetic hand control. In: 4th IEEE RAS and EMBS international conference on biomedical robotics and biomechatronics (BioRob). IEEE, pp 521–525
Van Erp JB, Toet A (2015) Social touch in human-computer interaction. Front Digital Humanities. Frontiers Media SA 2(2)

Chapter 4
The Next Generation CUFF: Smaller, Smarter, and More Efficient

Abstract This work introduces an enhanced version of the wearable haptic device, Clenching Upper-Limb Force Feedback (CUFF). This new design offers a smaller and more efficient device capable of delivering both skin stretch and grasping force feedback to the user's arm, making it highly suitable for prosthetic applications. In our research, we conducted thorough physical and perceptual tests with eleven able-bodied participants and carried out two key experiments involving a total of 32 participants. The CUFF device was used alongside the Pisa/IIT SoftHand robotic hand in both experiments. The first experiment focused on a discrimination task where participants differentiated between pairs of cylinders based on their size and perceived softness. In the second experiment, participants used the robotic hand to grasp various objects. After these tasks, participants provided feedback on their experiences with the device. The results from these experiments, along with participants' subjective evaluations, highlighted the effectiveness of the CUFF device. Thanks to its versatile design and functionality, this device shows great promise for a wide range of applications, including teleoperation, guidance, rehabilitation, and prosthetics.

4.1 Introduction

The sense of touch is an astonishingly powerful information generator. Different types of data from the external world are continuously processed by receptors distributed over the human body (Katz and Krueger 2013).

Wearable haptic system (WHS), which can be comfortably worn and integrated into everyday life (Pacchierotti et al. 2017), have shown promising results in fields such as robotic teleoperation (Ajoudani et al. 2014; Pacchierotti et al. 2015), rehabilitation (Bortone et al. 2017), and guidance (Culbertson et al. 2017). Furthermore, WHS can be integrated into the human body with minimal or no constraints on motion (Lieberman and Breazeal 2007), allowing for the study of human sensory perception

F. Barontini, *Wearable Haptic Devices for Realistic Scenario Applications*,
Springer Series on Touch and Haptic Systems,
https://doi.org/10.1007/978-3-031-70539-7_4

in a more ecological way. Typically, these devices can be applied to different parts of the human body, such as arms and forearms (Lieberman and Breazeal 2007), feet (Kim et al. 2006) and fingers (Leonardis et al. 2015) and are mainly thought to deliver tangential and normal force, as well as texture and softness cues, see e.g. Stanley and Kuchenbecker (2012) and Fani et al. (2017).

Even though the fingertip is generally intended as the privileged channel for discriminative touch (see Fani et al. 2017 for a complete overview of wearable haptic system on a fingertip), forearm and arm are often chosen as an effective site for haptic feedback delivery. This choice is mandatory in those cases when the hand location is not physically available, e.g. for amputees, or when constraints on feedback location impose a different placement of the device, e.g. for haptic guidance of people with visual impairments, who are usually required to have their hands free to properly interact and to understand the external environment (Barontini et al. 2020).

For these reasons, in literature, it is possible to find different systems that deliver force information to the arm and forearm, especially tangential force for haptic guidance and proprioception. In Yem et al. (2015) authors developed a wearable outer-covering haptic display (wOCHD) for hand motion, with two ball effectors to deform the skin and provide guiding information in four axes of motion. In Bark et al. (2010) authors presented a wearable haptic feedback device for rotational skin stretch to the hairy skin, which results as an effective means of providing information about a user's controlled joint or limb motions for motion training and similar applications. A similar solution is presented in Battaglia et al. (2019), and Aggravi et al. (2018) where a rotational skin stretch device is used to provide kinesthetic feedback from a prosthetic device. In Chinello et al. (2016) authors presented a lightweight bracelet with four servo motors to provide independent skin stretches at the palmar, dorsal, ulnar, and radial sides of the arm for haptic navigation tasks.

In addition to tangential and directional cues, pressure and force information also represents another key informative cue for task accomplishment, e.g. in prosthetics to convey grip information to amputees (Kim and Colgate 2012). In Meek et al. (1989) a grip pressure feedback for a myoelectrical controlled prosthetic arm was proposed. A servo-controlled "pusher" was mounted to the socket and pressed into the skin an amount proportional to the force in the terminal device. Such a method was named "extended physiologic taction" (EPT).

To the best of authors' knowledge, there is limited evidence of WHS capable of simultaneously delivering both normal and tangential forces, especially at the arm or forearm level.

In Minamizawa et al. (2007, 2010), authors presented a device that consists of two motors to convey both normal and tangential forces, thus simulating weight sensation. Analogously, a device—namely Hring—with two servo motors that move a belt placed in contact with the user's finger skin can provide both normal and shear force to the skin in Pacchierotti et al. (2016).

To address this gap, we present a wearable device, the Clenching Upper-limb Force Feedback (CUFF) device (Fig. 4.1), based on our preliminary work presented in the conference paper (Casini et al. 2015). The CUFF, worn on the forearm, is

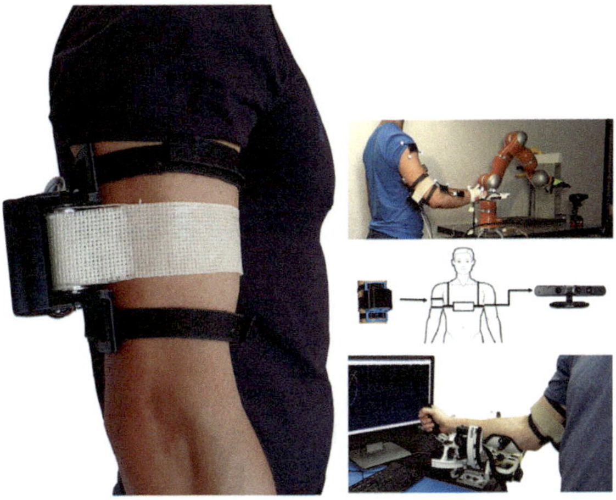

(a) (b)

Fig. 4.1 The CUFF Device applied on an subject's arm (**a**) and three possible applications (**b**): from the top, in a teleoperation setting, as assistive device in a guidance system, and as assistive device for rehabilitation

equipped with two independently controlled DC motors, enabling it to deliver both pressure and tangential cues in a fully wearable manner.

In comparison to our preliminary work (Casini et al. 2015), we have completely redesigned the device to enhance its wearability. This paper provides a thorough characterization of the system, reports on psychophysical experiments aiming to depict the perceptual workspace that the device can act on, and presents a series of exploratory experiments assessing the usability of the device in combination with a robotic hand.

4.2 Problem Definition

The problem of simultaneous rendering of multiple co-located stimuli over the same patch of skin is at the core of the design of cutaneous haptic devices.This is even more critical in wearable haptics due to the limited number of devices that can be comfortably worn, constrained by considerations of space and weight.

Looking at the corpuscles that contribute to various tactile modalities of the skin, they respond to a very broad range of different stimuli, including temperature variation, vibration, pressure, stretch, etc. In our design, we concentrate on the sensations of pressure and tangential stretch, as depicted in Fig. 4.2. We focus on these two aspects for the following reasons:

- Skin stretch can easily be associated with directional and navigation information (Gleeson et al. 2009);
- Pressure cues can effectively communicate grip force information for artificial limbs (Brown et al. 2015). This is of obvious relevance in prosthetics, but also in tele-operation (Aggravi et al. 2018) and guidance scenarios (Culbertson et al. 2017);

Fig. 4.2 Scheme of a arm, in section, with a distribution of normal pressure (left) and a distribution of tangential stress (right)

Skin stretch has been investigated as a way to convey proprioception feedback. There is evidence indicating that skin stretch plays a crucial role in conveying proprioception information of the hand and guidance information. In this sense, using skin stretch to convey proprioception of the prosthetic hands could lead to easier training for upper limb prosthesis users, since the feedback provided is felt in a way that is similar to a natural mechanism found in able-bodied limbs. Indeed, this type of feedback was shown to be an effective way to convey proprioception. In Bark et al. (2010) rotational skin stretch was proposed as an alternative to vibrotactile feedback for conveying proprioception and results showed it to be more effective. Regarding the second point, we were inspired by the prosthesis and blind people world. A common requirement from the amputees are able to operate prostheses without constant visual attention and proprioception and force feedback has been shown to improve targeting accuracy under non-sighted conditions (Fu and Santello 2018). Additionally, non-autonomous individuals with visual impairments are often guided by an accompanying person whose hand on their arm conveys directional information through stretch signals.

In Casini et al. (2015) we chose to focus on the skin surrounding the long trunks of the four limbs. This choice was motivated the need to keep hands and fingertips unencumbered, aligning with the approach often taken by consumer electronics. Numerous portable systems, such as smartwatches, smartphones, and sphygmomanometers, are designed to be strapped to the arm, forearm, thigh, or calf. In this paper we refine the design improving the previous work to obtain a device that is much more compact and lightweight, with a broader more in depth characterization of the system and its potential applications.

4.3 The CUFF 2.0

The device is the result of an iterative design process aimed at creating a lightweight, compact wearable device. Our focus was on the force perceived by users during its use, specifically investigating the range of force that ensures the system is pleasant and comfortable.

4.3.1 Design Improvement

Starting from the characterization of the first prototype (Casini et al. 2015), we know that the belt can elicit a maximum normal force of 22 N with a total current absorption equal to 1000 mA. It is worth noticing that the current absorption is the sum of the current absorbed by the two motors.

The first step was to study the force perceived by the subjects with the first prototype. We conducted an experimental campaign with 6 subjects (2 female, mean age 26 ± 2.1) with different arm sizes (28.90 ± 25.6 mm of circumference, measured with a tailor tape). The experiment aimed to estimate the maximum force we could exert on the user's arm, before reaching an unpleasant level. We defined the unpleasantness level as the level where the subjects feel the arm squeezed but without feeling pain.

During the experiment, the CUFF was worn on the participant's arm as in Fig. 4.1. The motors were controlled to move in opposite directions with a step profile, gradually increasing the current absorption by 20 mA every 2 s until the motor reached a stall condition. During the experiment we recorded the current absorbed by the motors, and the force elicited. The outcomes showed that the maximum force value we could exert was 17 N, corresponding to a current absorption of 800 mA. Keeping in mind these results, we chose the current motors, considering the required force value, the diameter for dimension reduction, and the reduction factor related to the length of the gear. The diameter was the main characteristic for dimension reduction, regarding the performed analysis we had a specific force value that had to be satisfied by the motor torque, and the reduction factor was chosen concerning the previous two characteristics and to the length of the gear itself. In Fig. 4.3 a render of the device (Fig. 4.3a and b) is shown, highlighting some details of its mechanical implementation. This release has a mass equal to $\approx$230 g, while its overall dimensions are $124 \times 70 \times 58$ mm (compared with the original design had a mass of $\approx$494 g

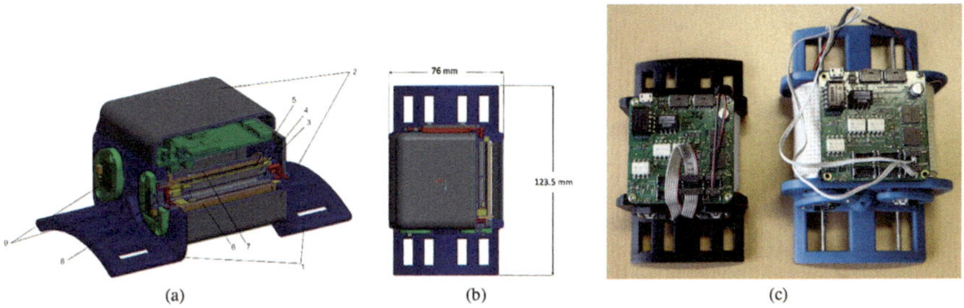

Fig. 4.3 Mechanical structure with all the components specified (**a**) and the size of the new prototype (**b**). In the figure is showed also the comparison between the old prototype, the one on the right side, and the new one on the left (**c**). In **a** it is possible to observe the (1) main frame, (2) cover, (3) Motor, (4) Motor custom housing, (5) Needle bearings, (6) Cylinders, (7) Fabric locker, (8) Encoder magnet and (9)Encoder

and overall dimensions 145 × 97 × 116 mm—see Fig. 4.3c). With the new design the volume of the device reduced of about 55% and the weight of about 56% (see Fig. 4.3c).

Figure 4.4 shows the effect of the two different stimuli on a Styrofoam cylinder, simulating an arm. The sequence from (a) to (c) shows the application of a normal force. In this case, the CUFF's motors rotate in opposite direction squeezing the cylinder. The stimulus elicits encoded like a force feedback stimulus. The sequences from (d) to (i) show the application of the tangential force. In this case, the motors rotate in the same direction, to the left or the right, generating a skin stretch suitable for a proprioception stimulus.

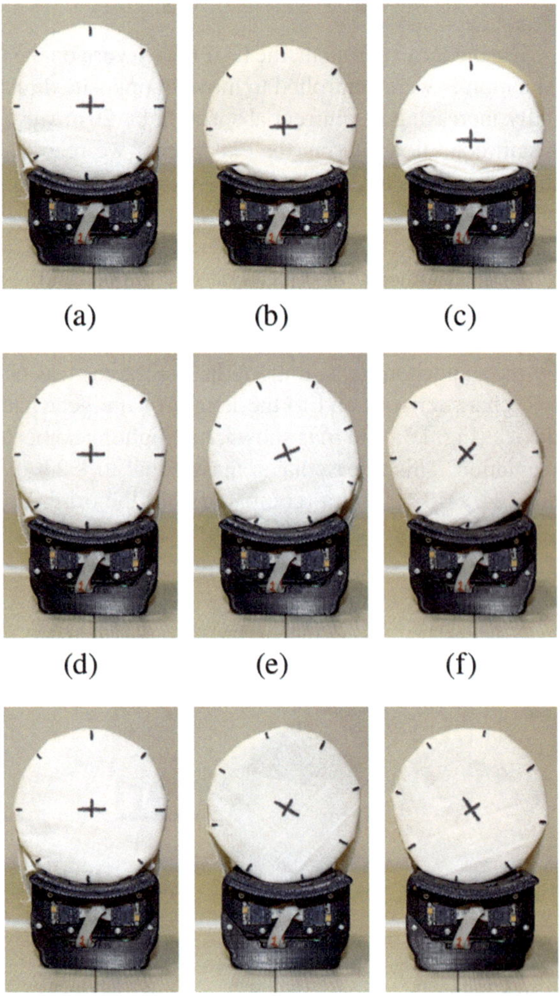

(a) (b) (c) (d) (e) (f) (g) (h) (i)

Fig. 4.4 Application of Normal and Tangential Force.The sequence from **a** to **c** shows the CUFF device applying increasing levels of normal force distribution to a styrofoam specimen. The sequence from **d** to **i** shows the CUFF device applying increasing levels of tangential force distribution to a styrofoam specimen

4.3.2 Mechanical Description

Figure 4.3a shows the device, which can be considered composed of three main subsystems: a structural frame (comprising parts 1, 2), the mechanical actuation units (3, 4, 5, 6) and the feedback interface (composed of a fabric belt not reported in the section view but visible in Fig. 4.1a). Each actuation unit is powered by a DCX10L (by Maxon Motor) with a two stage planetary gear-head with a 64:1 gear reduction. The motors (3) are attached to the main frame (1) through a custom housing (4), allowing a cylinder (6) to rotate around its main axis. Each actuation unit is sensed with a magnetic position sensor composed by the magnet (8), and its electronic board (9), (see Sect. 4.3.3). The fabric belt, used as main feedback interface with the human body, is fixed to the cylinders through fabric fixing plates. The belt is a band of non-elastic fabric (0.4 mm thick cotton 11—count aida cloth—10 squares per inch), covered on its internal side with a silicone layer, to allow a better grip on the user's skin (biphasic bio-compatible silicone C-MOL AK8 by Colorificio Rodoero SAS, Arenzano (GE), Italy). The device is fastened on the user's arm through a Velcro band.

4.3.3 Electronics and Software Description

The CUFF device is controlled using a custom-made electronic board, which is PSoC-based and uses the RS485 communication protocol. This board is integrated directly into the device and consists of two electrically separated main sections: the logic part and the power part. The logic part is responsible for reading the sensors and processing the signals, while the power part drives the motors. The motor driver serves as the interconnection point between these two sections. The board can be powered by a voltage range from 5 to 24 V, with all external logic ports operating at 3.3 V. The power output can be adjusted via firmware to meet the actuator's power requirements, regardless of the power source applied to the board. The device employs AS5045 Rotary Sensors from Austria Microsystems, which are contactless magnetic 12-bit rotary position sensors. These sensors provide precise angular measurements over a full rotation, with an angular resolution of approximately 0.0879°, equivalent to 4096 ticks per revolution. The CUFF is capable of functioning as a standalone device when connected directly to a robotic hand, offering force feedback, proprioception feedback, or a combination of both. Alternatively, it can be connected to a computer and controlled using its software, as well as C++ and Matlab©libraries, which are freely available online. This flexibility allows for straightforward and rapid integration into a wide range of applications. For further details and access to the software, additional information can be freely downloaded from the Natural Machine Motion Initiative website[1] (Della Santina et al. 2017).

[1] www.naturalmachinemotioninitiative.com.

4.4 Comprehensive Device Evaluation: Physical and Psychophysical Perspectives

To have a better knowledge of the device, we performed: a Physical characterization (see Sect. 4.4.1) of the new device, and a Psychophysical characterization (see Sect. 4.4.2) of the tangential displacement and normal force.

4.4.1 Physical Characterization

The goals of the process were: (1) to establish a relationship between the position of the motor and the normal force exerted by the fabric belt on the subject's arm (with a variable zero position), and (2) to determine the correlation between the exerted force and current absorbed by the motors.

A 3-axis force sensor, the ATI Gamma, was used to measure the normal force exerted on the subject's arm. We 3D printed structures with varied radii, aiming to replicate various human arm sizes. The selection of different radii was intentional to enhance the generalization of human arm dimensions. Starting from different zero positions, five cylinders were designed with the following dimension (80, 85, 90, 100, and 115 mm). Figure 4.5a and b show the finalized structure used for characterization and the manufactured cylinders, respectively.

4.4.1.1 Characterization Protocol and Control Modalities

The characterization was performed using a sequence of tighten-release movements of the CUFF. In this process, both motors were controlled to move in opposite direction to pull the fabric, avoiding any side-scrolling movements. The procedure was repeated ten times for each cylinder radius. At the beginning of any repetition, an auto-adjustment (hereinafter referred with pre-tensioning phase) of the fabric around the cylinder was performed. During this phase the fabric was tightened up to the stall condition of the two motors and then released up until the absorbed current was close to zero, guaranteeing the complete adhesion on the lateral surface of the cylinder. The pre-tensioning phase provided a zero-reference starting position for all characterization procedures, with the sensor showing $0.464\ \text{N} \pm 0.1793\ \text{N}$. Following the pre-tensioning phase, the tighten-release movement was performed.

We tested two control strategies: a current control, where we set a maximum current absortion of 1000 mA, and a position-current control, setting a maximum position of 800 motor ticks (corresponding to 17°). It is worth noticing that the position-current control consist of a double control loop (see Fig. 4.7). The position controller generates the reference current, and the current controller generates the PWM output, which powered the motor to generate the movement. This controller structure enables a double error check, one for the current and another one for the

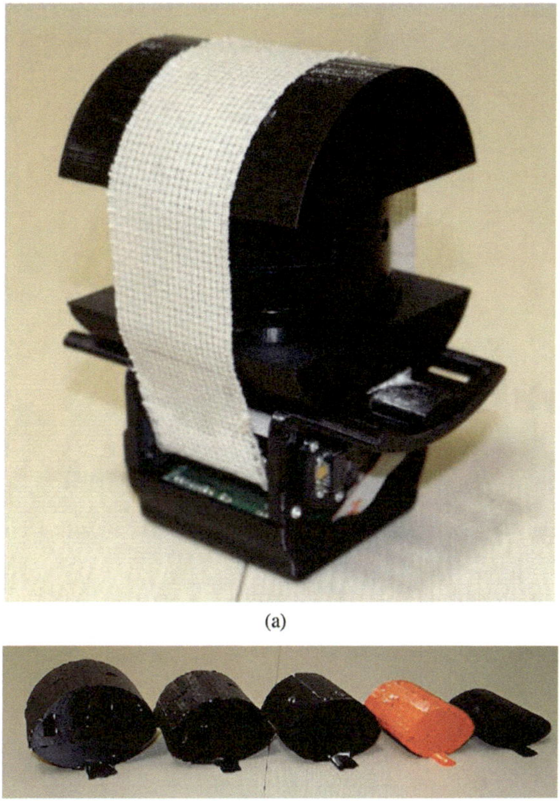
(a)
(b)

Fig. 4.5 Force sensor structure for characterization procedure mounted on the CUFF (**a**) and all the 3D-printed structures with the different radii (**b**)

position. During the tightening phase, the maximum value was reached in 60 s, and then the direction was reversed for other 60 s to return to the starting position. We recorded the current, the position of the motors and the normal forces exerted on the sensor. The control current exhibited a wide hysteris effect between the tighten-released phase (see Fig. 4.6), more so than the position-current control (Fig. 4.8a). Consequently, we opted to use the position-current control for psychophysical characterization and the experimental campaign with users.

4.4.1.2 Data Validation Procedure

To affirm the quality of the data acquired with the position-current control, we sought a function that could translate a desired force into a corresponding position command for the CUFF. The best fitting function for the dataset was a third order polynomial function:

$$P_D = f(F_D) = 0.1138 F_D^3 - 5.204 F_D^2 + 89.22 F_D \tag{4.1}$$

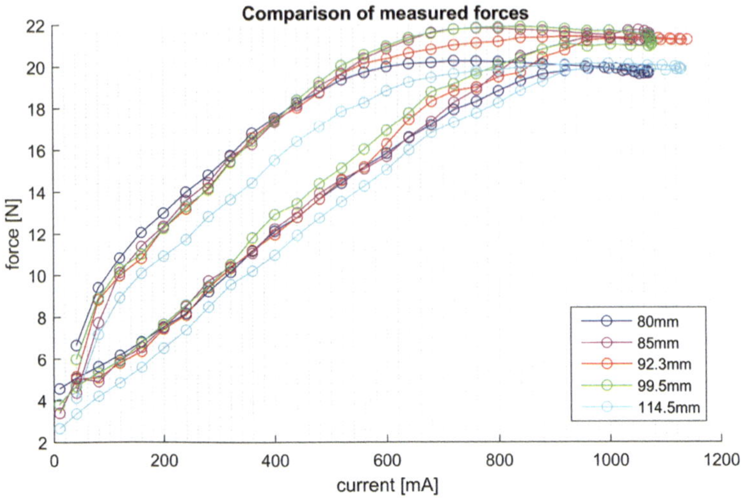

Fig. 4.6 Characterization curve obtained with non-elastic fabric band and current control

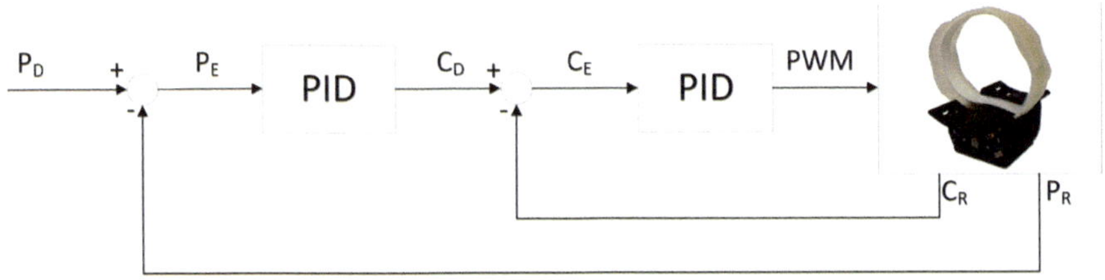

Fig. 4.7 CUFF double control loop—External position loop and internal current loop

Where P_D is the desired motor position and F_D is the desired force (Adjusted R-square of 0.9335).

We validated the fitting (see Fig. 4.8b) in another series of experiments. We used the function $f(F_D)$ to derive the position values, command the CUFF, and measure the forces we obtained. The Root Mean Square Error (RMSE) between the theoretical force values and measured ones was 1.32 N across all experimental trials for all the radii (see Fig. 4.9).

4.4.2 Psychophysical Analysis

The goal of the Psychophysical characterization was to estimate user's ability to discriminate the normal force and the tangential displacement delivered with the CUFF. We used the method of the constant stimuli to find the Just Noticeable Difference (JND), as defined in Gescheider (2013), i.e., the minimum amount of

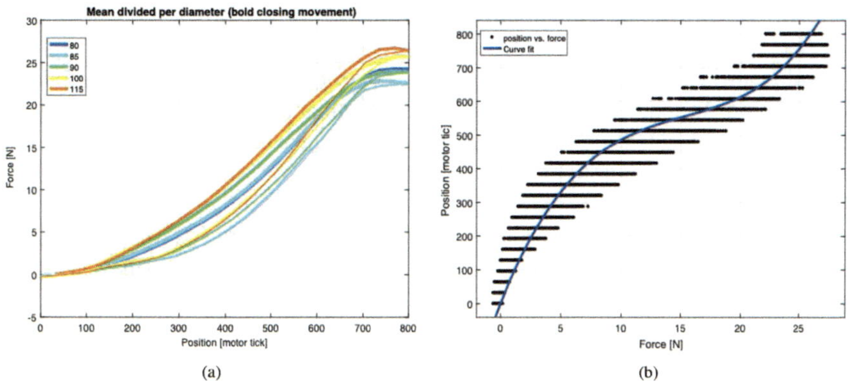

Fig. 4.8 In **a** the characterization data are showed, in the plot the mean values for each diameter for the tighten movement (thicker lines) and releasing one are reported; **b** shows the data set (in black) and the obtained fitting curve (in blue)

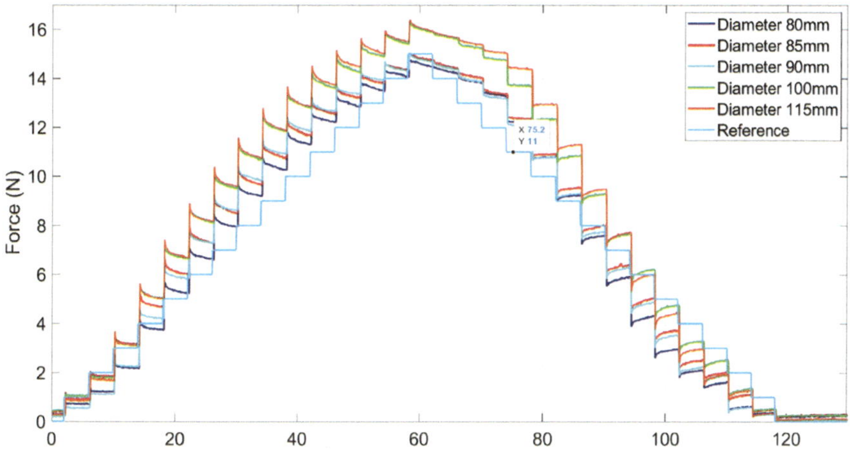

Fig. 4.9 Validation experiment—In orange the theoretical force values, while the other colors refer to the measured force values for different cylinder radii

displacement/force that can elicit a different perception in users with respect to a reference stimulus. We considered to use the device in the two operating modes depicted in Fig. 4.2. Eleven right-handed able-bodied participants (7 females, mean age $\pm$ SD: 26.64 $\pm$ 8.86) gave their informed consent to participate in the experiment. No one had any physical limitation which would have affected the experimental outcomes. The average arm diameter (using a cylindrical geometrical approximation) for participants was 80.3 $\pm$ 7.7 mm (mean $\pm$ SD).

4.4.2.1 Tangential Displacement Perception

Stimulus and Procedure: Participants were comfortably seated, with their forearm resting on a desk and, wearing the cuff on their dominant arm, see Fig. 4.1. During the experiments, subjects wore headphones with white noise to prevent auditory cues; in addition, they were required to not look at the device during the experiments. Before these experiments, the CUFF was applied to the arm of the subject and we performed a pre-tensioning and rescaling procedure. The pre-tension was performed to bring the fabric in contact with the subject's skin as previously described with the sensorized cylinders. The new position of the motors was used as zero reference position for all the trials. The rescaling was performed to rescale the positions obtained with the fitting function to the arm biomechanical structure. More specifically, the rescaling was computed commanding to the motor a position much higher than the maximum one that can be effectively reached by the CUFF motors (due to the current limits). We know that the maximum reachable position corresponding to the maximum current absorption provides on average a force of 25 N on the sensor. The effective position of CUFF motors obtained in this way was then associated with the value of 25 N and all the other points of the characterization were re-scaled accordingly.

Participants received paired stimuli, where each stimulus corresponded to the tangential displacement generated by the rotation of the CUFF's motors in the same direction. They were asked to indicate which stimulus in each pair was perceive as higher. Each pair comprised a reference stimulus (RS) and a comparison stimulus (CS), presented in a pseudo-random order. The displacement in the reference stimulus was consistently 17.91 mm, while in the comparison stimulus, it was selected from five discrete and equally spaced values (5.97, 11.94, 17.91, 23.88, and 29.85 mm). A single trial consisted of : the first stimulus, an inter-stimulus interval of 2 s, and the second stimulus followed by the subject's response. We considered two experimental blocks corresponding to the direction of the tangential motion (i.e. rightward or leftward). Each block consisted of 100 trials. The binary responses of the participants were recorded.

Data Analysis: We modelled the responses of each volunteer using the psychometric function

$$\Phi^{-1}\left[P\left(Y_j = 1\right)\right] \sim \beta_0 + \beta_1 x_j \tag{4.2}$$

where $\left[P\left(Y_j = 1\right)\right]$ is the probability that, in trial j, the participant reported a larger stimulus in the comparison than in the reference stimulus, Φ^{-1} is the probit transformation of the response probability (i.e., the inverse function of the cumulative normal distribution), and x_j is the value of the comparison stimulus.

We extended the psychometric function to the whole pool of participants through a Generalized Linear Mixed Model (GLMM) (Moscatelli et al. 2012), which enables to take into account the analysis of clustered data. In our case a cluster is the collection of repeated responses in several participants. GLMM is a hierarchical model that provides both predictions on the experimental effects, which are assumed to be systematic across participants, and an estimate of the variability between participants.

For each experimental condition, we estimated the Just Noticeable Difference (JND), i.e. the amount of stimulus change for a difference to be noticeable and the Point of Subjective Equality (PSE), i.e. the stimulus value yielding a response probability of 0.5. The JND provides an estimate of the precision of the response: the lower is the JND, the easier is to discriminate two stimuli of different lengths. The PSE estimate the accuracy of the response: the response is accurate if the PSE is non-significantly different from the reference stimulus. For each parameter, we estimated the 95% confidence intervals (CIs) using the bootstrap method described in Moscatelli et al. (2012).

Results: The GLMM fit to the data is illustrated in Fig. 4.10c. The JND was equal to 2.91 mm (95% CIs: 2.61–3.21 mm) for rightward moving stimuli and 2.70 mm (95% CIs: 2.40–2.96 mm) for leftward moving stimuli. The difference in JND between the two conditions was not statistically significant (difference: 0.22 mm; 95% CIs of the difference from −0.19 to 0.63 mm). The PSE was equal to 17.42 mm (95% CIs: 16.89–17.95 mm) for rightward moving stimuli and 18.28 mm (95% CIs: 17.76–18.83 mm) for leftward moving stimuli. This means that the response was accurate for rightward moving stimuli, and slightly biased for leftwards (although the bias was smaller than 1 mm). The 95% CI of the difference between PSEs does not include zero (difference: −0.85 mm; 95% CIs of the difference from −1.54 to 0.18 mm). This small bias must be further investigated. A possible explanation could be related to the non-linearity of human arm biomechanics.

4.4.2.2 Normal Force Perception

Stimulus and Procedure: The experimental setup and procedure were the same as for the tangential displacement task. Participants experienced paired stimuli and they were asked to indicate which stimulus in the pair produced a higher normal force. The force values were equal to 9 N in the RS and was pseudo-randomly chosen among five stimuli (3, 6, 9, 12, 15 N) in CS. The stimulus duration was 1 s, while the inter-stimulus interval was 2 s.

Data Analysis: Data were analyzed with modalities similar to the one of tangential displacement task.

Results: The estimated JND was 2.21 N (95% CIs: 1.95–2.47 N). The estimated PSE was 9.75 N (95% CIs: 9.38–10.11 mm). That is, the responses were precise (the participants were able to discriminate a force difference as small as 2 N). As for the accuracy, there was a small difference between the PSE and the reference stimulus (smaller than 1 N). In Fig. 4.11 it is possible to see the GLMM fit.

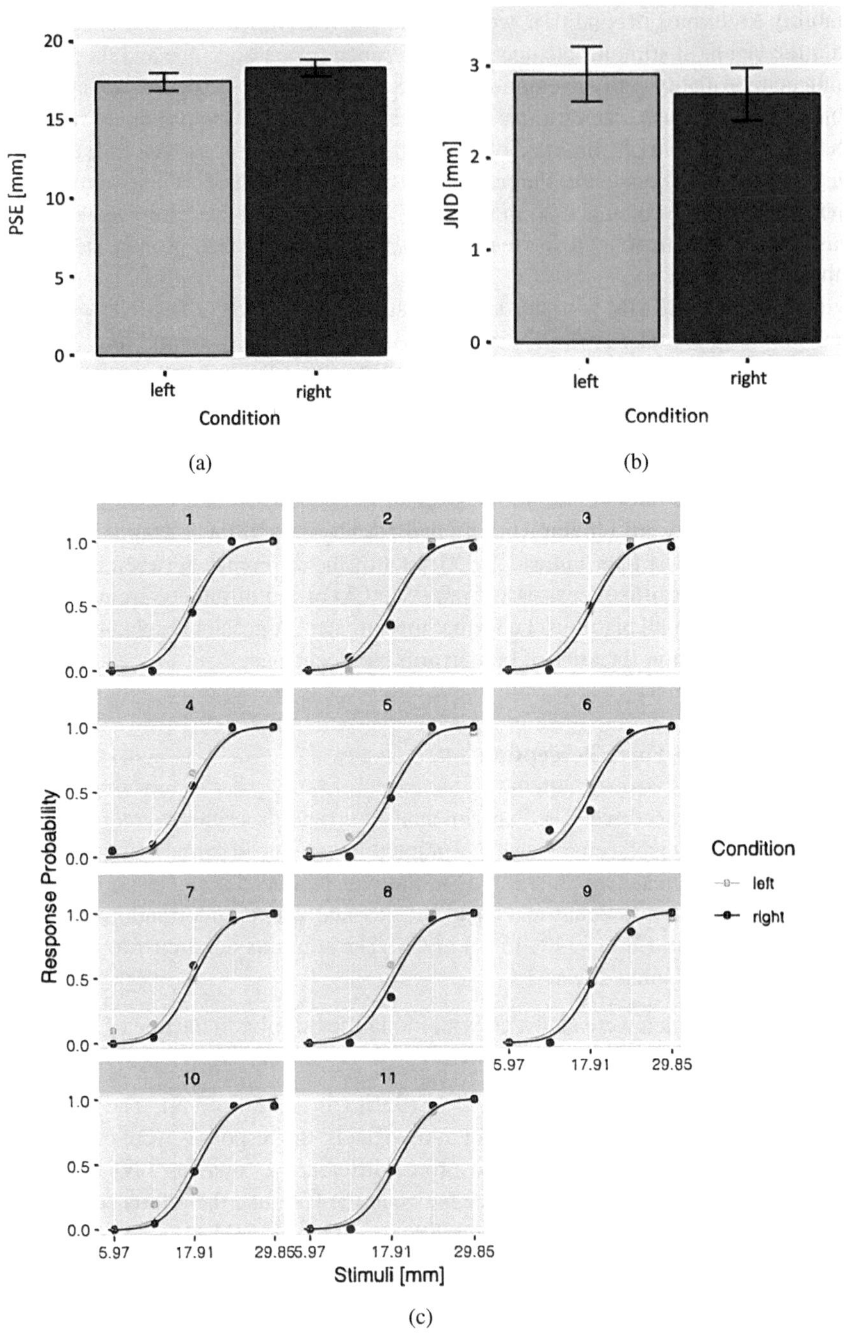

Fig. 4.10 Results of the displacement test. **a** and **b** present bar plots for JND and PSE under two conditions. **c** displays raw data and GLMM predictions for the right and left conditions across all subjects. Each box represents raw data and model predictions for individual participants (labeled 1–11)

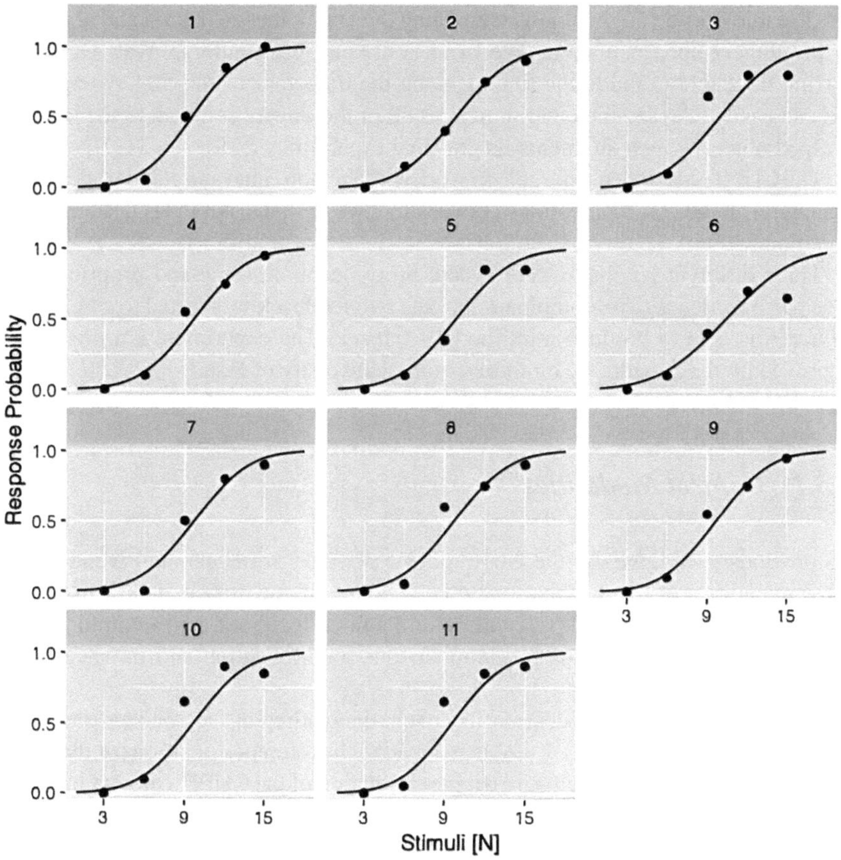

Fig. 4.11 Results of the GLMM fit for the force test across all subjects. Each box represents both raw data and model predictions for individual participants, labeled from 1 to 11

4.5 Explorative Experiments Involving Human Participants

A series of experiments were performed to test the device in two of its possible applications, as wearable haptic feedback device for prosthetic applications, and for remote control of avatars. The two possible degrees of freedom of the device can be fully used in this application, enabling the generation of different stimuli connected to different perceptions. Other examples of applications of the CUFF are discussed in Sect. 4.2. For these experiments, the CUFF was used with an underactuated soft robotic hand, the Pisa/IIT SoftHand (Catalano et al. 2012). The SoftHand (SH) is a kinematic synergy-inspired hand with 19 degrees of freedom (DOFs). Elastic ligaments allow the phalanges to deform and ensure their return to the initial position. A single cable runs through all joints, enabling the simultaneous flexion and abduction of the fingers upon actuation. The hand is driven by a single 15 Watt DC motor driving the cable to guide the fingers along the trajectory of the first synergy. The SH can easily adapt to the environment as all of the fingers conform to the shape of grasped objects which augments its grasping capabilities.

The CUFF is designed to be able to work with the SoftHand as a standalone device, managing all the communications between the two systems. No PC is needed for the communication.

Three different feedback were tested: proprioception, force and proprioception plus force feedback. The combination of the two feedback was tested to assess if the device was able to produce multimodal feedback. The experimental protocol was approved by the Bioethics Committee of the University of Pisa.

4.5.1 Control Modalities

As previously said, we use the CUFF device with the SoftHand, and we identified three modalities of stimuli, that could be generate by the device: a skin stretch for the proprioception feedback, a squeeze for the force feedback, and a combined skin stretch and squeeze stimulus for simultaneous proprioception and force feedback delivery.

In the context of proprioception feedback, the reading of the encoder positioned on the SoftHand's motor (p_{SHmeas}) is mapped to the reference position for the CUFF motors (p_{Cref}). Each hand closure triggers a rotation of the CUFF's motors in the left direction (conventionally chosen), generating a skin stretch stimulus. The operating range of the SH comes from 0 ticks (hand completely open) to 19000 ticks (hand completely closed). Instead, concerning the force feedback, the residual current of the SoftHand, rc_{SHmeas}, is mapped into opposite reference position for the CUFF motors, rc_{Cref}, producing a squeezing stimulus on the user's arm. It is worthy notice that the residual current is defined as the difference between the estimated current and the real current absorbed by the SoftHand motor. In combined feedback stimuli, both the motor position and residual current data from the SoftHand are used to generate a coordinate movement of stretch and squeeze on the user's arm. The tactile stimulation of the CUFF device directly depends on the information received from the SoftHand. Specifically, the encoder's measurement can provide valuable insights into whether the hand is grasping an object and thereby offer clues about the object's dimensions. Additionally, the residual current data can indicate the softness of an object, where a higher residual current suggests a harder object, potentially providing information about its fragility. The CUFF device receives the information directly from the SoftHand via RS485 protocol at a frequency range of 1 kHz.

Mapping information from the robotic hand into the feedback device involves associating each hand value with a corresponding stimulus delivered by the device.

Mapping the information coming from the robotic hand into feedback device, means associate to each value of the hand a corresponding stimulus delivered by the device. At the beginning of our investigation, we chose to implement a linear function to associate each position and the residual current value from the SoftHand to the CUFF device's reference position. This linear mapping operate under two constrain: $p_{SHmeas} = 0$ corresponded to $p_{Cref} = 0$ and $p_{SHmeas} = p_{SHmax}$ to $p_{Cref} = p_{Cmax}$; where p_{SHmax} is the maximum value of the SoftHand encoder and p_{Cmax} is the maximum value of the CUFF encoders. Analogous constrains were applied to the linear mapping of the residual current: $rc_{SHmeas} = 0$ corresponded to $rc_{Cref} = 0$ and $rc_{SHmeas} = rc_{SHmax}$ to $rc_{Cref} = rc_{Cmax}$. Therefore, the reference position is computed as:

$$p_{Cref} = \left(\frac{p_{SHmeas}}{p_{SHmax}}\right) p_{Cmax} \tag{4.3}$$

$$rc_{Cref} = p_{SHmeas} * gain \tag{4.4}$$

respectively for the proprioception feedback (4.3), and for the force feedback (4.4), with a $gain = 0.4$ (heuristically found).

After an initial series of experiment, we decided that a non-linear mapping might be a better solution for the task of size and force discrimination. Consequently, we choose to implement an exponential mapping for the proprioception feedback, and logarithmic mapping for the force feedback for the subsequent experiments. We used the same constrains used for the linear mapping, $p_{SHmeas} = 0$ corresponded to $p_{Cref} = 0$, but we imposed that $p_{SHmeas} = \frac{2}{3} p_{SHmax}$ was mapped to $p_{Cref} = \frac{1}{3} p_{Cmax}$. The new reference positions were computed as follows:

$$p_{Cref} = \alpha e^{-\beta\left(\frac{p_{SHmeas}}{p_{SHmax}}\right)} p_{Cmax} \tag{4.5}$$

$$rc_{Cref} = \gamma \log\left(1 - \delta \frac{rc_{SH_m eas}}{rc_{SHmax}}\right) rc_{Cmax} \tag{4.6}$$

The constrains were imposed by choosing the appropriate values for the constants: $\alpha = 0.1547$, $\beta = 1.944$, $\gamma = 0.9510$ and $\delta = -0.3317$.

For the task of discriminating combined information (size plus force), we initially used linear mapping, and subsequently utilized the exponential and logarithmic mapping (see Fig. 4.12) for size and force respectively.

4.5.2 *Passive Evaluation: The Discrimination Task*

We tested the effectiveness of our system with a total of 32 participants. The main goal of these experiments was to verify if our feedback device could be a potential solution to translate informations about the size and the softness of an object in a proprioception and force stimuli. We investigated also, the possibility to give a

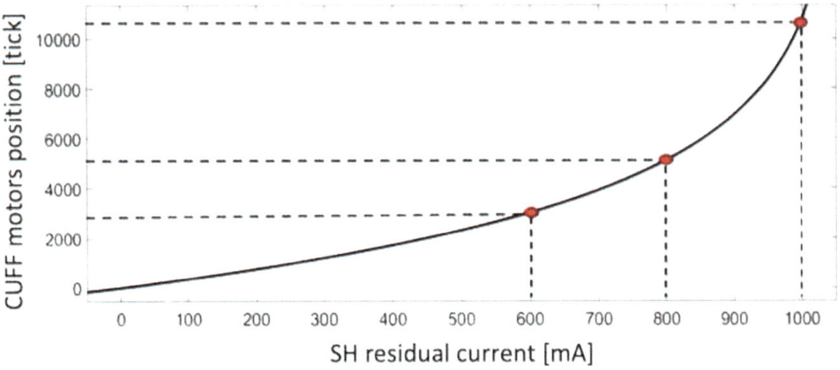

Fig. 4.12 The logarithmic mapping used to control the CUFF's motors in the force feedback modality

combine information of proprioception and force at the same time. None of the participants had any physical or mental limitation which could affect the experimental outcomes.

4.5.2.1 Participants

Thirty-two able-bodied subjects all right-handed gave their informed consent to participate to the discrimination task. Sixteen participants (5 females, mean age $\pm$ SD: 27.68 ± 2.6), perform the task of testing a linear mapping (Experiment A) between the CUFF and the SoftHand for all the modalities. Sixteen subjects (4 female, mean age $\pm$ SD: 28.87 ± 2.6), accomplished the task testing the logarithmic mapping for the grasping force modality, and the exponential mapping for the proprioception modality (Experiment B). Both the mappings were used for the combined information modality.

4.5.2.2 Setup and Procedure

The experimental design consisted of a training phase and three discrimination tasks: (1) proprioception modality, (2) force modality, and (3) combined stimuli modality. During the experiment, the participants were seated comfortably with the CUFF worn on the right arm, and the forearm resting on the table. They were blindfolded and wore headphones with pink noise to eliminate auditory cues generated by the rotation of CUFF motors. In each discrimination task, participants perceived stimuli on their arm generated by the feedback device operating in a playback modality. This entails the utilization of prerecorded signals from the SoftHand, captured during the grasping phase of various cylinders, to control the rotation of the CUFF motor. The

CUFF was connected to a computer via a USB cable, and the recorded signals were transmitted to the subjects through an execution program. We selected six cylinders of different materials: three rigid cylinders 3D printed in Rapid Prototyping (ABS) material, and three soft cylinders made of foam rubber. Moreover, we choose three sizes of diameters (40, 60, 80 mm). The cylinders were used for the discrimination test of size, softness and the combined information (size plus softness).

For the proprioception discrimination task, we recorded the p_{SHmeas} and rc_{SHmeas}, including instances when the hand was fully closed around the object. Each cylinder was oriented horizontally in the palm of the SoftHand and was grasped ten times. For the softness discrimination task, we utilized a 60 mm diameter cylinder with an internal 3-axis force sensor (ATI Gamma) to measure the force elicited by the SoftHand during three different levels of closures. We opted for final closure position of 15000, 16500, and 18000 ticks of the hand. In this case too, the cylinder was positioned horizontally in the palm of the robotic hand, we recorded the residual current value, and the procedure was repeated ten times for each closure level. It is noteworthy that this strategy was adopted to ensure participants remained focused on the stimuli elicited by the device.

During the 15-min training period, subjects were made familiar with the dimension and material of the cylinders. They were instructed to associate the stimulus generated by the CUFF with the corresponding cylinders. The order of execution for the proprioception and grasping force modality was randomized and counterbalanced between the subjects. The combined stimuli modality was always conducted as third task, to allow participants to first gain a clear understanding of the proprioception and force stimulus individually.

Proprioception Modality: Participants were asked to identify the sliding stimuli associated with the size of three rigid cylinders and the stimulus correlated with the hand closing without an object (i.e. close hand without object, ⌀40 mm, ⌀60, ⌀80 mm).

During each trial, the subjects perceived on their arm a sliding movement of CUFF's belt to the left. The belt moved less, when the SoftHand grasped a larger cylinder, on the other hand the belt moved more when the SoftHand closed around small cylinder or without any object. Subjects were presented sixteen pairs of stimuli and asked to determine whether the second stimulus was smaller, larger or of equal sliding compared to the first stimulus. To avoid artifacts due to temporal errors, each pair of stimuli was presented two times, one with the first element of the pair being presented as first and another one with the first element of the pair presented as second in temporal order, for a total of twenty trials in random order.

Force Modality: In this task, the participants had to recognize the intensity level of three squeeze stimuli and the stimulus originated by the SoftHand closing without an object.

The subject perceived a squeeze stimulus elicit by the belt around their arm for each trial. The belt squeezed the arm more when the SoftHand grasped a rigid object, and less when it grasped a softer object. For this modality, we utilized the residual current values obtain during the recording session of the sensorised cylinder (close hand without object, 15000 ticks, 16500 ticks, 18000 ticks). Like the proprioception modality, participants were presented sixteen pairs of grasp stimuli and asked to determine if the second stimulus was greater, lesser or of equal intensity compared to the first. Each pair was presented twice, as in the proprioception modality, yielding a total of twenty trials in random order.

Combined Stimuli Modality: Participants were asked to recognize the combined stimuli of proprioception plus grasping force. Participants first perceived the leftward slide of the belt, immediately followed by the squeeze of the arm. Using these two pieces of information, participants should be able to distinguish the size and softness of the cylinder.

As previously discussed, the belt would move less, if the hand grasped a larger cylinder, and it would squeeze harder when the hand grasped a rigid cylinder. For this task, we used the signals recorded from the two sets of cylinders (i.e. close hand without object, rigid ⌀40 mm, rigid ⌀60 mm, rigid ⌀80 mm, and soft ⌀40 mm, soft ⌀60 mm, soft ⌀80 mm). Subjects were presented with forty-nine pairs of combined stimuli. Participants were asked two questions to discern the dimension of size and stiffness. We asked whether the second stimulus was larger, smaller, or of equal displacement compared to the first, aiming to identify the dimension. Additionally, we asked if the second stimulus was larger, smaller, or of equal intensity compared to the first, aiming to identify differences in stiffness. Each pair was presented twice to avoid artifacts in the results, resulting in a total of fifty-six trials presented in random order.

4.5.2.3 Data Collection

To evaluate the performance in the discrimination task we recorded the number of correct answers for all the discrimination tasks (see Fig. 4.13), and conducted a subjective quantitative questionnaire evaluated on a 7-point Likert scale. The questionnaire consists of a series of inquiries related to the system and the experimental task. Participants were asked to assign scores ranging from 1 *totally disagree* to 7 *totally agree*. This approach represents a common procedure to evaluate devices for assistive robotics and Human-Robot Interaction (Prattichizzo et al. 2013). The questions delivered to the subjects are listed in Table 4.1. The survey included questions for all the participants of the discrimination task (Q1 to Q21) and additional questions for the participants who accomplished the fragile object task (Q22 to Q38). Questions Q1 to Q5 referred to the wearability of the CUFF device, questions Q6 to Q8 were related to visual and auditory distraction. Questions Q9 to Q21 investigated the intuitiveness of sensory substitution, the sensation provided by the device, and the performance.

Questions Q22 to Q38 explored the use of the CUFF device in conjunction with the SoftHand, investigating the performance, stimuli usefulness, and task fatigue.

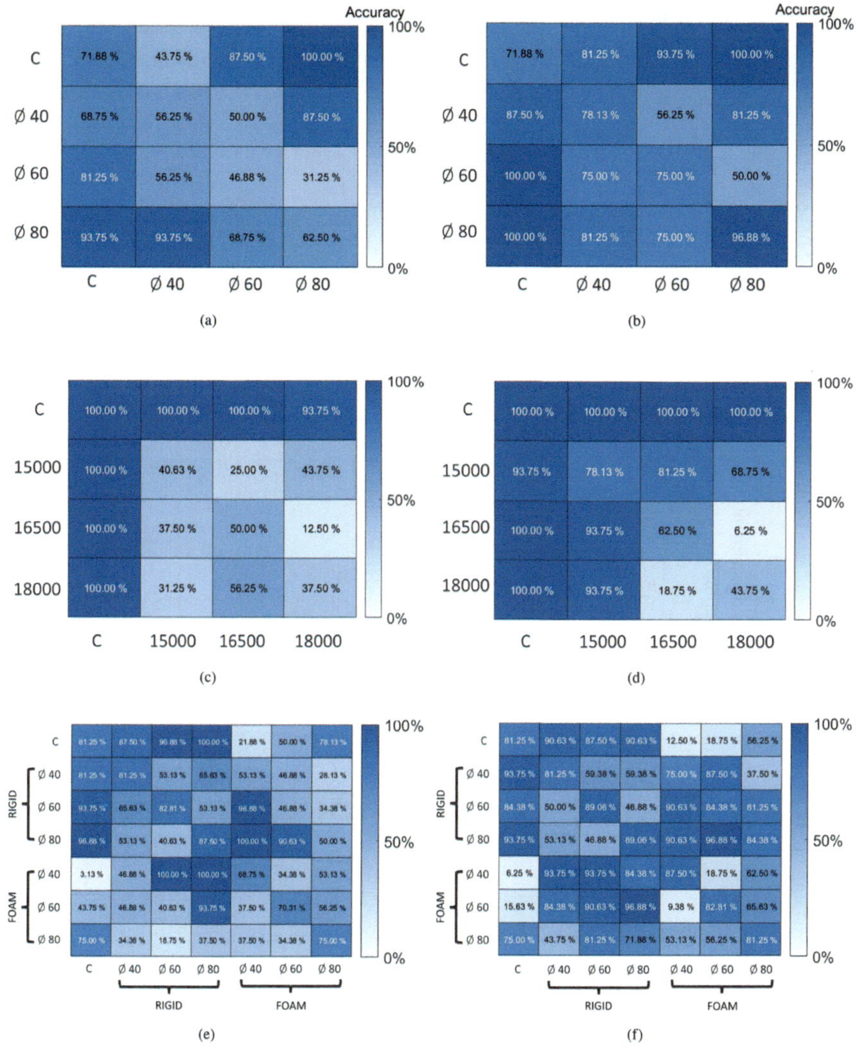

Fig. 4.13 These pictures depict the results obtained from the discrimination tasks. The first row displays the outcomes of the proprioception modality using linear mapping (**a**) and exponential mapping (**b**). The values on the horizontal and vertical dimension represent the diameter of the cylinders used during the experiment. The second row presents the results of the force modality using linear mapping (**c**) and exponential mapping (**d**). The vertical and horizontal values represent the SoftHand closures expressed in ticks. Finally, the third row illustrates the results of the combined stimuli modality using linear mapping (**e**) and exponential plus logarithmic mapping (**f**). The vertical and horizontal values represent the diameters, expressed in mm, of the rigid and foam cylinders used during the experiment

Table 4.1 These statements were assessed by participants in Experiments A and B using a 7-point Likert scale (1: Strongly Disagree, 7: Strongly Agree). In Experiment A, linear mapping was employed, whereas in Experiment B, participants utilized exponential and logarithmic mapping. Median values and standard deviations across all individuals are presented. The asterisks highlight the questions that demonstrated statistical significance ($p < 0.05$) based on the Wilcoxon rank-sum test analysis

	Questions	Participants experiment A (Linear mapping)		Participants experiment B (Exponential/logarithmic mapping)		p-value
		Median	Iqr.	Median	Iqr.	
Q1	I felt hampered by the cutaneous device	2	0.5	2	1	
Q2	The belt produced pain during long time usage	2	1	1	1	
Q3	The noise generated by the actuators interfered with the haptic perception	2	2	1	0.5	
Q4	The sensation provided by the CUFF on the arm felt pleasant	5	2.5	5	1.5	
Q5	The sensation provided by the CUFF on the arm felt unpleasant	2	2.5	2.5	2	
Q6	During the discrimination task the CUFF and the cylinder were out of my visual field	7	0	7	0	
Q7	During the discrimination task, I was able to see the cylinder or the CUFF device	1	0	1	0	
Q8*	During the discrimination task, I was able to hear the sounds made by the motors of the CUFF	1	3	1	0	0.0038
Q9*	It has been easy to discriminate the cylinder size using the CUFF in proprioception modality	5	2	6	1	0.024
Q10	Discriminating the cylinder size without looking at them was very difficult using the CUFF	2	1.5	2	1	
Q11	The proprioception feedback (sliding movement) was clear	5	1	6	0.5	
Q12	I was able to understand the position of the hand also after a distraction, perception retention	6	1	4	2.5	
Q13	It has been easy to discriminate the cylinder softness using the CUFF in force modality	5	1.5	5	1	

(continued)

Table 4.1 (continued)

	Questions	Participants experiment A (Linear mapping)		Participants experiment B (Exponential/logarithmic mapping)		p-value
		Median	Iqr.	Median	Iqr.	
Q14	Discriminating the cylinder softness without looking at them was very difficult using the CUFF	2.5	2	2	1	
Q15	I was able to distinguish different levels of force through the cutaneous device	6	1	6	1	
Q16	The maximum force elicit pain	1	1	1	0	
Q17	It has been easy to discriminate the cylinder softness using the CUFF in combined modality	6	1	6	1.5	
Q18*	It has been easy to discriminate the cylinder size using the CUFF in combined modality	2.5	1	5	1.5	0.0131
Q19	Discriminate the cylinder size without looking at them was very difficult using the CUFF	2.5	1	2	1	
Q20	Discriminate the cylinder softness without looking at them was very difficult using the CUFF	2.5	1	3	1	
Q21*	The two stimuli were clearly distinguishable	6	0	5	1	0.0047
Q22	It has been easy to use the SoftHand together with the CUFF	/	/	6	0.5	
Q23	I was feeling uncomfortable while using the SoftHand together with the CUFF	/	/	2	1	
Q24	It was easy to move the virtual eggs while receiving the proprioception feedback by the CUFF	/	/	6	1.5	
Q25	It was easy to move the virtual eggs while receiving the force feedback by the CUFF	/	/	6	0.5	
Q26	It was easy to move the virtual eggs while receiving the proprioception feedback by the CUFF during the combined modality	/	/	6	1.5	

(continued)

Table 4.1 (continued)

	Questions	Participants experiment A (Linear mapping)		Participants experiment B (Exponential/logarithmic mapping)		p-value
		Median	Iqr.	Median	Iqr.	
Q27	It was easy to move the virtual eggs while receiving the force feedback by the CUFF during the combined modality	/	/	6	0.5	
Q28	I had feeling of performing better while receiving force feedback or proprioception feedback by the CUFF with respect to no feedback	/	/	6	1	
Q29	I had the feeling of performing better when I was not receiving any feedback by the CUFF	/	/	2	1	
Q30	It has been easy to discriminate the cylinder size without any feedback	/	/	1	0	
Q31	It has been very difficult to discriminate the cylinder size without any feedback	/	/	7	0	
Q32	It has been easy to discriminate the cylinder softness without any feedback	/	/	1	1	
Q33	It has been very difficult to discriminate the cylinder softness without any feedback	/	/	7	0	
Q34	It has been easy to discriminate the cylinder size without any feedback during the combined modality	/	/	1	0	
Q35	It has been very difficult to discriminate the cylinder size without any feedback during the combined modality	/	/	7	0	
Q36	It has been easy to discriminate the cylinder softness without any feedback during the combined modality	/	/	1	0	
Q37	It has been very difficult to discriminate the cylinder softness without any feedback during the combined modality	/	/	7	0.5	
Q38	At the end of the experiment I felt tired	/	/	2	1	

*The asterisks highlight the questions that demonstarted statistical significance ($p < 0.05$) based on the Wilcoxon rank-sum test analysis

4.5.2.4 Results Discrimination Task

Figure 4.13 illustrates the results of discrimination experiments conducted with linear mapping (Fig. 4.13a, c, and e) and exponential/logarithmic mapping (Fig. 4.13b, d, and f). The overall accuracy for each experiment conducted with linear mapping is 68.75%, 64.26%, and 61.73% for proprioception, force, and the combined modality, respectively. In contrast, for exponential and logarithmic mapping, the total accuracy is 81.45% for proprioception, 77.54% for force, and 68.72% for the combined modality. Furthermore, for the combined modality, the accuracy in correctly recognizing both stimuli was calculated. The results show an accuracy of 55.86% (for proprioception) and 67.06% (for force) with linear mapping, while with exponential/logarithmic mapping, the accuracy is 64.66% and 72.77% for proprioception and force, respectively. Additionally, the collected data underwent analysis using the Wilcoxon-Mann Whitney test. Significant values (p-value <0.05) were found for all three conditions: 0.007 for proprioception, 0.0129 for force, and 0.0158 for the combined modality. The same analysis was performed for individual responses within the combined modality, yielding p-values of 0.021 for proprioception and 0.05 for force. Holm-Bonferroni correction was applied to the found values, and even after correction, the values remained significant (0.0021, 0.0258, 0.0158), including for individual conditions within the combined modality (0.0042, 0.0495). Table 4.1 reports the qualitative questionnaire with the corresponding Likert scale evaluation. Questions Q1 to Q3, related to the wearability of the device, showed a positive rate with a median value of 2, 2, 2 and an interquartile range of 1, 1.22 and 0.99. Questions Q6 to Q12, concerning the proprioception feedback discrimination task obtained a positive evaluation. Particularly, question Q11 had a median value of 5 and an interquartile range of 1 (for the linear mapping), and a median value of 6, and a interquartile range 0.5 (for the logarithmic/exponential mapping). Questions Q13 to Q16, which investigated the force feedback discrimination task, showed similar positive results. The performance of the task (Q13) had a median value of 5 and an interquartile range of 1.5 (linear mapping), and a median and interquartile range value of 5 and 1 respectively (logarithmic/exponential mapping). Of note is question Q16, regarding the maximum force elicited by the device, which presented a median value of 1, 1 and a interquartile range of 1, 0 for linear and logaritmic mapping respectively. Questions Q17 to Q21, regarding the combined stimuli feedback task, also displayed an encouraging response. Moreover, we analyzed the qualitative questionnaire using the Wilcoxon rank-sum test. The results revealed significant differences between the two experimental modalities, linear (Experiment A) and exponential/logarithmic (Experiment B), for questions Q8, Q9, Q18, and Q21, with p-values of 0.0038, 0.024, 0.0131, and 0.0047, respectively.

4.5.3 Handling Evaluation: The Fragile Object Task

We tested the effectiveness of our system with a total of 16 participants. The primary goal of these experiments was to determine if our feedback device could be a useful solution during manipulation and grasping actions. Similar to the discrimination task, we investigated the single modalities (force and proprioception individually) as well as the combined feedback of proprioception and force. None of the participants had any physical or mental limitations that could affect the experimental outcomes.

4.5.3.1 Participants

Sixteen subjects (4 Female, mean age $\pm$ SD: 28.87 $\pm$ 2.6), all right-handed, gave their informed consent to participate to the experiment.

4.5.3.2 Setup and Procedure

We introduced an additional task where the subjects controlled the SoftHand using electro-myographic sensors (EMGs). The CUFF device was worn on their arm and the EMG sensors were positioned on the Flexor Digitorum Superficialis (FDS) and on the Extensor Digitorum Communis (EDC), and secured by an elastic band. The SoftHand was connected to a gravity compensator, the SaeboMAS, to support participant's arm during the experiment. The participants stood in front of a desk, with their arm placed on the gravity compensator. They wore headphones with pink noise to avoid auditory cue, and glasses with covered lenses to prevent a clear view of the boxes.

Participants were asked to perform a task similar to a Box and Block test: in one minute they had to move as many fragile objects as possible from a point A to a point B, overcoming an obstacle without breaking the objects see Fig. 4.14. This task was used in Engels et al. (2018) assesses grasping efficiency. The subjects performed the task in four modalities: (1) no feedback, (2) proprioception feedback, (3) force feedback, and (4) combined stimuli feedback. Each participant repeated the task four times, one for each modality. The order of modality execution was randomized.

The fragile object is a box made of paper, dimension 50 $\times$ 50 mm, with two lateral sides open and, a fragile fuse placed in the middle of the other two side. The fragile fuse is a piece of pasta, fragile enough to break with a pinch grasp. We tested three types of spaghetti with a diameter of 1.1, 1.5 and, 2 mm, and we characterized a piece of spaghetti 50 mm long, to compression with an Instron instrument. We repeated the test 15 times for each diameter, and recorded the force corresponding at the breaking point.

For the experiment, we selected the spaghetto with a diameter of 1.5 mm, which had a breaking force of approximately 10 N.

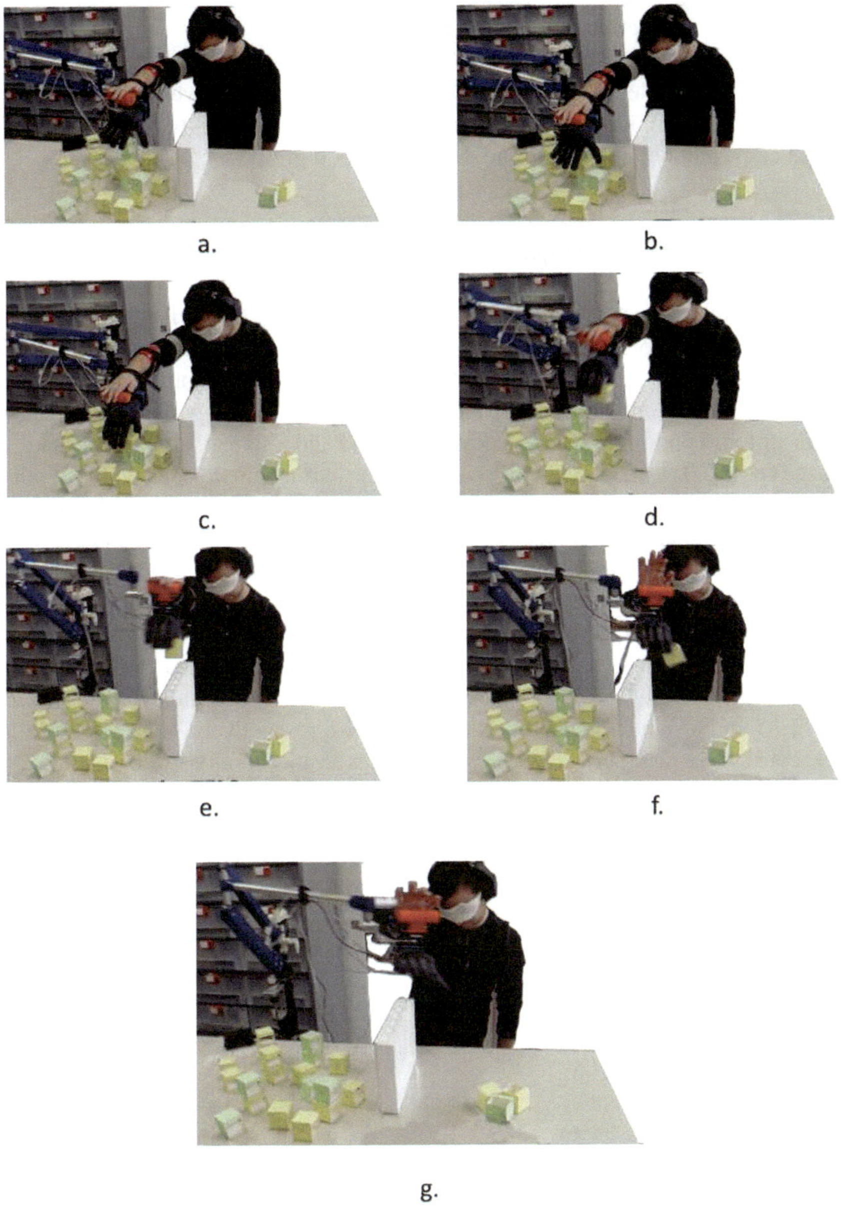

Fig. 4.14 A sequence of the experimental setup used during the manipulation evaluation

4.5.3.3 Data Collection

Regarding the fragile object task, in addition to the qualitative survey, we also recorded for each modality of execution: total number of eggs moved, number of broken eggs, number of regrips, and the wrong/good grasp (see Table 4.2).

4.5.3.4 Results Fragile Object Task

Table 4.2 summarizes the outcomes of the fragile object task. The median values for the overall number of moved fragile boxes were 10, 11, 10, and 9, with interquartile ranges of 6, 5, 4, and 5.5 for the 'No feedback,' 'Proprioception,' 'Force,' and 'Combined stimuli' modalities, respectively. On average, less than 1 box was broken across all four conditions. The Friedman test indicated a statistically significant difference ($p < 0.05$) among the four conditions for 'wrong grasp' data, with a p-value of 0.0159. Additionally, for this dataset, we applied the Wilcoxon signed-rank test, revealing a significant difference between the 'No feedback' condition and the 'Proprioception' condition (p-value = 0.0039), as well as between the 'No feedback' condition and the 'Proprioception + Force feedback' condition (p-value = 0.0293). Moreover, subjects who took part in the fragile object task, underwent also the questions Q22 to Q38. For question Q22 and Q23, which addressed the wearability and the use of the device in conjunction with the SoftHand, the mean values were 6.06, 1.87 and a standard deviation of 0.68 and 0.8 respectively. Questions Q28 to Q29, concerning task performance, were rated with mean values of 6.18 and 1.87 and a standard deviation of 0.75 and 0.71. Questions Q30 to Q38, which probed the usefulness of the stimuli elicit by the device, received a positive rate. For question Q38, regarding the fatigue, mean and standard deviation values were 2.56 and 1.41.

Table 4.2 The table shows the median and interquartile range to accomplish the virtual eggs tasks for each different experimental modality

	No feedback		Proprioception		Force		Combined stimuli	
	Median	Iqr.	Median	Iqr.	Median	Iqr.	Median	Iqr.
Broken egg	0	1.5	0	0	0	1	0	1.5
Regrip	1	2	0	1	0	1.5	0	1.5
Wrong grasp	1	1.5	0.5	1	0	1	0	1
Good grasp	9	5	10	4.5	9.5	5.5	9	5.5
Total egg	10	6	11	5	10	4	9	5.5

4.6 Discussion

The psychophysical characterization results, confirmed by the experimental validation with able-bodied subjects, demonstrated the efficacy of the CUFF device in conveying proprioception and force feedback stimuli. The tangential displacement showed no statistically significant difference between the leftward and rightward sliding stimulus. Furthermore, subjects were capable of discerning a force difference as small as 2 N.

The results of the discrimination task, obtained using linear mapping, indicated that participants could correctly recognize pairs of cylinders, although they struggled with cylinders of similar diameters (e.g., diameter 60 mm, 80 mm). This challenge can be attributed to the minor differences in encoder values, which, when linearly mapped, result in small movements, less than 1000 ticks, of the CUFF motors. In the force feedback discrimination task, participants accurately discerned the difference between a closed hand and all other closure levels. However, distinguishing all other pairs proved challenging. It's noteworthy to highlight the difficulty participants experienced in perceiving stimuli individually and in the combined modality.

Analyzing the percentage of correct responses separately for proprioception and force in the combined modality revealed that subjects found it more challenging to distinguish the size of the object. This challenge is attributed to the combined perception of the two stimuli, where skin-stretch, in some instances, could be overshadowed by a distinct arm squeeze, confusing the subjects.

Statistical analysis with the Wilcoxon-Mann Whitney test clearly highlighted how the use of exponential/logarithmic mapping helped subjects improve task performance. Applying the Holm-Bonferroni correction, the p-values remained significant (<0.05). The results confirmed that using a different mapping from the linear one could enhance the performance and efficacy of the stimuli delivered by the CUFF device on the user's arm. These findings were further validated by the qualitative questionnaire results (see Figs. 4.14, 4.15 and 4.16). Participants generally found the device easy to wear and use, even when combined with the robotic hand (Q1, Q2, Q22, and Q23). The sensory substitution was considered intuitive, and the stimuli assisted in discriminating the cylinder size and softness (Q9, Q13, Q18). More specifically, the overall impression of the participants was that the stimuli were clear and easy to recognize separately (Q21), although some difficulties were experienced when receiving the combination of both stimuli. Those who performed the virtual eggs task felt they performed better with feedback from the CUFF compared to no feedback (Q28, Q29).

The results from the fragile object task imply that participants felt a significant improvement in performance when receiving stimuli from the CUFF compared to no feedback (Q28, Q29). The Wilcoxon signed-rank test results (p-value $= 0.0039$, p-value $= 0.0293$) support this perception, indicating that feedback significantly enhanced subjects' abilities to execute more accurate and precise grasps, despite the number of total objects moved being nearly the same. Despite this positive outcome, there were no noteworthy distinctions among the various feedback modalities,

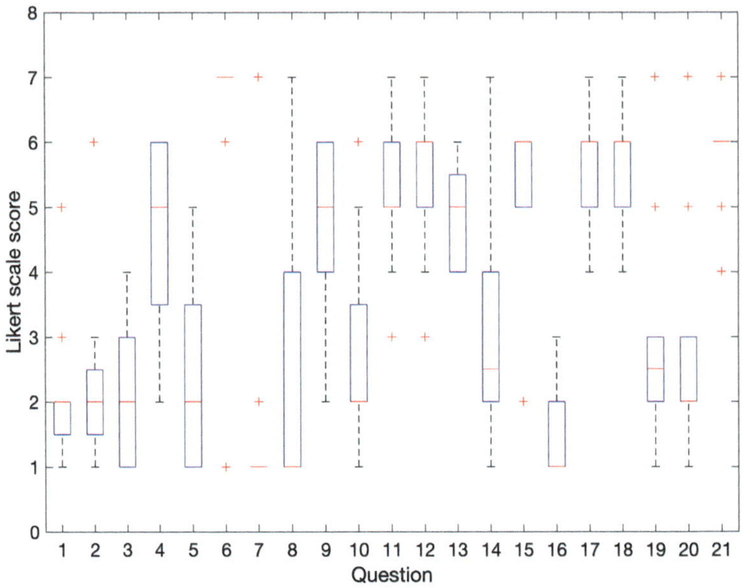

Fig. 4.15 The boxplot shows the results of the Likert Scale of the participants who accomplished the Experiment A. Center lines show the medians; box limits indicate the 25th and 75th percentiles; whiskers extend 1.5 times the interquartile range from the 25th and 75th percentiles. n = 16 participants to the Experiment A

suggesting an absence of a clear preference for one feedback type over another. Consistently, the count of broken eggs remained zero across all conditions, as depicted in Table 4.2. While this outcome is promising, it may not comprehensively showcase the true effectiveness of the feedback. The absence of a significant difference in task performance with or without feedback could be attributed to the task's simplicity, which did not demand a high level of dexterous manipulation. Subsequent experiments could delve into whether utilizing different fragile objects (e.g., heavier ones) would yield enhanced subject performance with feedback compared to its absence. Interestingly, participants reported that the feedback made the task more intuitive (Q31, Q32), helped them understand the hand's state (Q34, Q35), and improved their immersion (Q37). Additionally, participants mentioned that the presence of feedback did not augment cognitive load when performing the task (Q38). For a complete overview of the qualitative questionnaire regarding the fragile objects task, see Fig. 4.17. Despite this, the participants' preference for feedback suggests that it could provide a more intuitive and immersive control experience. This is a promising indicator for the potential benefits of incorporating sensory feedback into prosthetic devices.

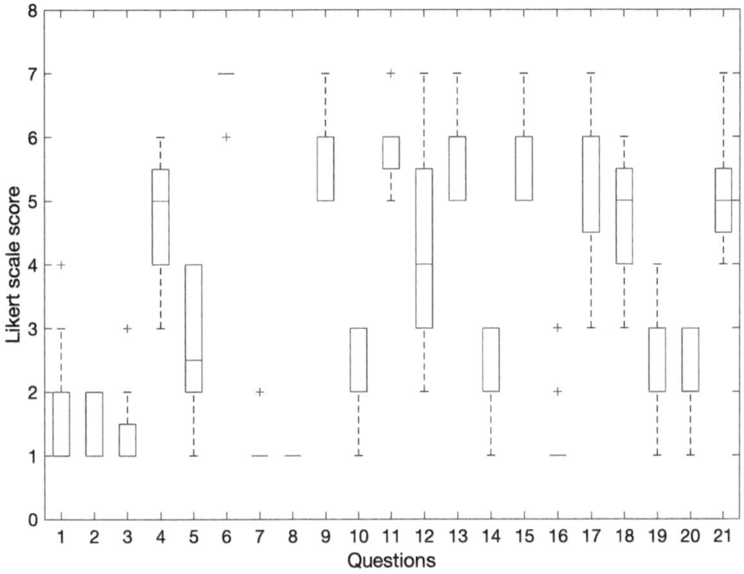

Fig. 4.16 The boxplot shows the results of the Likert Scale for the participants of Experiment B. Center lines show the medians; box limits indicate the 25th and 75th percentiles; whiskers extend 1.5 times the interquartile range from the 25th and 75th percentiles. n = 16 participants to the Experiment B

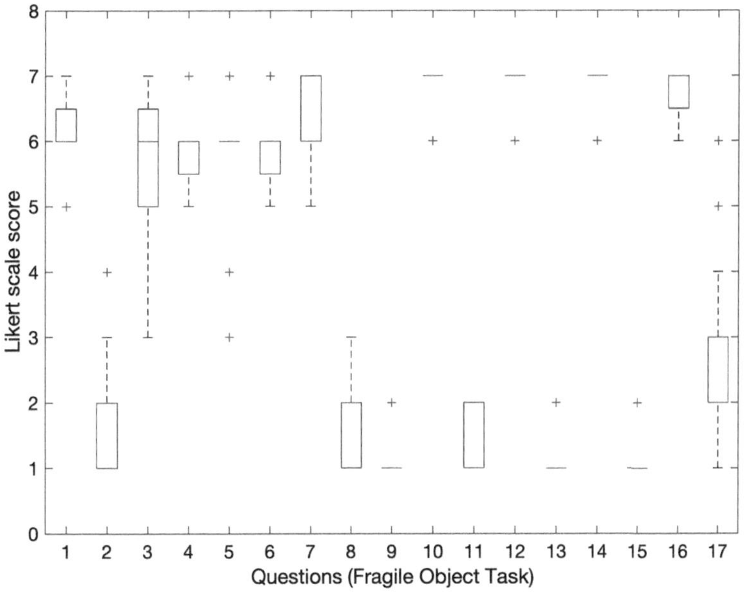

Fig. 4.17 The boxplot shows the results of the Likert Scale for the participants who accomplished the Fragile Object task. Center lines show the medians; box limits indicate the 25th and 75th percentiles; whiskers extend 1.5 times the interquartile range from the 25th and 75th percentiles. n = 16 participants to the Experiment B (Fragile object task)

4.7 Conclusions

We introduced the CUFF device, a wearable haptic device capable to convey proprioception and force feedback stimuli, either individually or in combination. The device underwent a rigorous validation process through physical and psychophysical characterization, and was subsequently tested in an experimental campaign involving thirty-two able-bodied participants. The psychometric results demonstrated that the stimuli generated by the device are well-perceived by the users when worn on the upper arm. This was true for both normal force and tangential sliding stimuli, with no discernible difference between the two sliding directions. Our experiments focused more on the CUFF device used in tandem with the SoftHand in size and grasping force discrimination tasks. The participants were able to use the feedback from the device to distinguish the size, the grasping force, and the softness of virtualized cylinders. Survey responses indicated that the device was intuitive and effective. Notably, the logarithmic and exponential mapping approach exhibiting better results compared to the linear mapping method. Even if the experiments were performed only by able-bodied subjects, the encouraging results suggest that further investigation with limb loss participants could yield valuable insights. It would be of interest to conduct psychophysical tests by instructing participants to don the device on their non-dominant arm, aiming to investigate potential disparities between the dominant and non-dominant arms. Future developments involve applying the device at various points along the arm to assess the impact of different locations on force distribution. Finally, as highlighted within the paper, the current functionality of the device is limited to providing directional cues related to right/left rotations. Future developments will also encompass the capability to deliver up/down stimuli.

Acknowledgements The authors would like to thank Nicoletta Colella for her help in the characterization process and Mattia Poggiani for his useful help in the system optimization.

References

Aggravi M, Pausé F, Giordano PR, Pacchierotti C (2018) Design and evaluation of a wearable haptic device for skin stretch, pressure, and vibrotactile stimuli. IEEE Robot Automat Lett 3(3):2166–2173

Ajoudani A, Godfrey SB, Bianchi M, Catalano MG, Grioli G, Tsagarakis N, Bicchi A (2014) Exploring teleimpedance and tactile feedback for intuitive control of the pisa/iit softhand. IEEE Trans Haptics 7(2):203–215

Bark K, Wheeler J, Shull P, Savall J, Cutkosky M (2010) Rotational skin stretch feedback: a wearable haptic display for motion. IEEE Trans Haptics 3(3):166–176

Barontini F, Catalano MG, Pallottino L, Leporini B, Bianchi M (2020) Integrating wearable haptics and obstacle avoidance for the visually impaired in indoor navigation: a user-centered approach. IEEE Trans Haptics 14(1):109–122

Battaglia E, Clark JP, Bianchi M, Catalano MG, Bicchi A, O'Malley MK (2019) Skin stretch haptic feedback to convey closure information in anthropomorphic, under-actuated upper limb soft prostheses. IEEE Trans Haptics 12(4):508–520

Bortone I, Leonardis D, Solazzi M, Procopio C, Crecchi A, Briscese L, Andre P, Bonfiglio L, Frisoli A (2017) Serious game and wearable haptic devices for neuro motor rehabilitation of children with cerebral palsy. In: Converging clinical and engineering research on neurorehabilitation II: proceedings of the 3rd International Conference on NeuroRehabilitation (ICNR2016), October 18–21, 2016, Segovia, Spain, pp 443–447. Springer International Publishing

Brown JD, Paek A, Syed M, O'Malley MK, Shewokis PA, Contreras-Vidal JL, Davis AJ, Gillespie RB (2015) An exploration of grip force regulation with a low-impedance myoelectric prosthesis featuring referred haptic feedback. J Neuroeng Rehabil 12:1–17

Casini S, Morvidoni M, Bianchi M, Catalano M, Grioli G, Bicchi A (2015) Design and realization of the cuff-clenching upper-limb force feedback wearable device for distributed mechano-tactile stimulation of normal and tangential skin forces. In: IEEE/RSJ international conference on intelligent robots and systems (IROS), pp 1186–1193. IEEE

Catalano MG, Grioli G, Serio A, Farnioli E, Piazza C, Bicchi A (2012) Adaptive synergies for a humanoid robot hand. In: 12th IEEE-RAS international conference on humanoid robots (Humanoids 2012), pp 7–14. IEEE

Chinello F, Pacchierotti C, Tsagarakis NG, Prattichizzo D (2016) Design of a wearable skin stretch cutaneous device for the upper limb. In: IEEE Haptics Symposium (HAPTICS), pp 14–20. IEEE

Culbertson H, Walker JM, Raitor M, Okamura AM (2017) WAVES: a wearable asymmetric vibration excitation system for presenting three-dimensional translation and rotation cues. In: Proceedings of the 2017 CHI conference on human factors in computing systems, pp 4972–4982

Della Santina C, Piazza C, Gasparri GM, Bonilla M, Catalano MG, Grioli G, Garabini M, Bicchi A (2017) The quest for natural machine motion: an open platform to fast-prototyping articulated soft robots. IEEE Robot & Autom Mag 24(1):48–56

Engels LF, Cappello L, Cipriani C (2018) Digital extensions with bi-axial fingertip sensors for supplementary tactile feedback studies. In: 2018 7th IEEE international conference on biomedical robotics and biomechatronics (Biorob), pp 1199–1204. IEEE

Fani S, Ciotti S, Battaglia E, Moscatelli A, Bianchi M (2017) W-FYD: a wearable fabric-based display for haptic multi-cue delivery and tactile augmented reality. IEEE Trans Haptics 11(2):304–316

Fu Q, Santello M (2018) Improving fine control of grasping force during hand-object interactions for a soft synergy-inspired myoelectric prosthetic hand. Front Neurorobot 11:71

Gescheider GA (2013) Psychophysics: the fundamentals. Psychology Press

Gleeson BT, Horschel SK, Provancher WR (2009) Communication of direction through lateral skin stretch at the fingertip. In: World Haptics 2009-Third Joint EuroHaptics conference and symposium on haptic interfaces for virtual environment and teleoperator systems, pp 172–177. IEEE

Katz D, Krueger LE (2013) The world of touch. Psychology Press

Kim K, Colgate JE (2012) Haptic feedback enhances grip force control of sEMG-controlled prosthetic hands in targeted reinnervation amputees. IEEE Trans Neural Syst Rehabil Eng 20(6):798–805

Kim H, Seo C, Lee J, Ryu J, Yu SB, Lee S (2006) Vibrotactile display for driving safety information. In: 2006 IEEE intelligent transportation systems conference, pp 573–577

Leonardis D, Solazzi M, Bortone I, Frisoli A (2015) A wearable fingertip haptic device with 3 DoF asymmetric 3-RSR kinematics. In: 2015 IEEE World Haptics Conference (WHC), pp 388–393. IEEE

Lieberman J, Breazeal C (2007) TIKL: Development of a wearable vibrotactile feedback suit for improved human motor learning. IEEE Trans Robot 23(5):919–926

Meek SG, Jacobsen SC, Goulding PP (1989) Extended physiologic taction: design and evaluation of a proportional force feedback system. J Rehabil Res Dev 26(3):53–62

Minamizawa K, Fukamachi S, Kajimoto H, Kawakami N, Tachi S (2007) Gravity grabber: wearable haptic display to present virtual mass sensation. In: ACM SIGGRAPH 2007 emerging technologies, pp 8-es

Minamizawa K, Prattichizzo D, Tachi S (2010) Simplified design of haptic display by extending one-point kinesthetic feedback to multipoint tactile feedback. In: 2010 IEEE Haptics symposium, pp 257–260. IEEE

Moscatelli A, Mezzetti M, Lacquaniti F (2012) Modeling psychophysical data at the population-level: the generalized linear mixed model. J Vis 12(11):26–26

Pacchierotti C, Meli L, Chinello F, Malvezzi M, Prattichizzo D (2015) Cutaneous haptic feedback to ensure the stability of robotic teleoperation systems. Int J Robot Res 34(14):1773–1787

Pacchierotti C, Sinclair S, Solazzi M, Frisoli A, Hayward V, Prattichizzo D (2017) Wearable haptic systems for the fingertip and the hand: taxonomy, review, and perspectives. IEEE Trans Haptics 10(4):580–600

Pacchierotti C, Salvietti G, Hussain I, Meli L, Prattichizzo D (2016) The hRing: a wearable haptic device to avoid occlusions in hand tracking. In: 2016 IEEE Haptics symposium (HAPTICS), pp 134–139. IEEE

Prattichizzo D, Chinello F, Pacchierotti C, Malvezzi M (2013) Towards wearability in fingertip haptics: a 3-dof wearable device for cutaneous force feedback. IEEE Trans Haptics 6(4):506–516

Stanley AA, Kuchenbecker KJ (2012) Evaluation of tactile feedback methods for wrist rotation guidance. IEEE Trans Haptics 5(3):240–251

Yem V, Otsuki M, Kuzuoka H (2015) Development of wearable outer-covering haptic display using ball effector for hand motion guidance. In: Haptic interaction: perception, devices and applications, pp 85–89

Chapter 5
User-Centered Indoor Navigation System for the Visually Impaired

Abstract In this work, we propose a novel indoor navigation system based on wearable haptic technologies, aimed at increasing the autonomy and improving the quality of life for blind individuals. Despite promising research outcomes in the field of technological travel aids, these solutions have seen limited acceptance in real-world applications, due in part to the insufficient involvement of end-users in the conceptual and design phases. Our proposed system was developed with continuous feedback from visually impaired persons. It consists of an RGB-D camera, a processing unit that computes visual information for obstacle avoidance, and the CUFF device, which provides normal and tangential force cues for guidance in unfamiliar indoor environments. Experiments with blindfolded subjects and visually impaired participants demonstrate that our system could be an effective support during indoor navigation and a viable tool for training blind individuals in the use of travel aids.

5.1 Introducation and Motivation

All over the world, 285 millions of individuals are estimated to be visually impaired. More specifically, 39 millions are totally blind, while 246 have low visual capabilities (WHO 2014). According to the European Forum Against Blindness (EFAB 2013), more than 200,000 Italians are blind and around 6 millions suffer from eye diseases. Blindness dramatically limits the quality of life of these people and their families, especially in terms of autonomous navigation and privacy. Furthermore, it also represents a considerable economic burden for the society: looking at the sole case of Italy, the annual costs related to blindness are over 2 billions Euros (EFAB 2013). It is not hence surprising that a lot of effort has been dedicated to improve life conditions of blind individuals and their relatives, especially for what concerns the enhancement of blind people autonomy. Under this regard, mobility assistance for visually-impaired people represents a very challenging task, since it requires the reconstruction and intuitive delivery of spatial information on the environment, for enabling a safe navigation. Specially-trained guide dogs and white canes represent the most widespread systems for navigation and obstacle avoidance. However, they

F. Barontini, *Wearable Haptic Devices for Realistic Scenario Applications*, Springer Series on Touch and Haptic Systems,
https://doi.org/10.1007/978-3-031-70539-7_5

come with important drawbacks due to the extensive training required, the need for occupancy of one user's hand, and the reduced amount of spatial information they can provide to the user (mainly limited e.g. to low obstacles).

In recent years, different technological solutions have been proposed to increase blind people walking autonomy. These systems are generally referred to as Electronic Travel Aids (ETAs) (Dakopoulos and Bourbakis 2009). ETAs can surrogate spatial information on position and obstacle location and deliver it to the users via sensory substitution, relying on auditory or haptic cues. ETAs can be portable or wearable, the latter ones are usually preferred since they can be used in conjunction with classical travel aids, e.g. the white cane. For a complete review on these topics please refer e.g. to Velazquez (2010), SungMin (2016).

Looking at the sensory substitution approaches, the practical usability of the acoustical feedback is often limited for mobility applications. This because it might interfere and reduce user's auditory ability—which is usually augmented in blind people, disturbing posture and equilibrium control and severely affecting social interactions (Moller et al. 2009; Ando 2015). For the reasons above, tactile feedback seems to represent a more natural manner to convey navigational information. This observation has been strenghtened with the avenue of wearable haptic systems (WHSs), which have gained an increasingly important role for the delivery of haptic cues to the human wearers, with special focus on the cutaneous ones (see Pacchierotti et al. 2017 for a review on this topic).

Generally speaking, the haptic exploration mode can be distinguished in two different classes: active and passive (Smith 1997; Rodriguez 2019; Zelek 2005). In the active haptic exploration, people actively use their hands to gather haptic information during the interaction with the device. On the contrary, in the passive haptic exploration the hands of the users are still, while the device haptically elicits touch-related perceptions on the fingers, palm or dorsum (or at other body locations in case of wearable systems). Among the different types of portable solutions, which are based on the latter type of exploration, it is worth mentioning guide dog robotic systems and smart canes for indoor and outdoor navigation. The GuideCane (Borenstein and Ulrich 1997) is a wheeled system, which can detect the obstacle and steers around it; the user feels this steering action on the hand and follows the robot. In Wei et al. (2014) the authors developed a smart rope interactive system connected with the user hand. The NavCane (Meshram et al. 2019) is a smart cane device able to deliver priority information about obstacles in the path. The priority information is transmitted to the user using tactile and auditory communication methods. In Wang and Kuchenbecker (2012), Ando (2015), Wahab et al. (2011), the authors developed different smart cane devices that can convey information using tactile stimulation. In Kassim et al. (2016), Saegusa et al. (2011) the information regarding the environment and the presence of obstacles, which is gathered through different types of sensors, is conveyed to the blind user via vibration or auditory cues.

All these devices can be also categorized as portable because they require the hand of the user to continuously hold them. In addition, wearable haptic solutions have been also developed to convey tactile navigation cues. According to the passive haptic exploration mode, which mainly rely on vibration stimulation (Keyes et al.

2015) and, eventually integrated in gloves designed for haptic feedback delivery, see e.g. Keyes et al. (2015). Indeed, vibrotactile stimulation represents the most common tactile modality used in ETAs, which has been also used in wearable belts and devices for the wrist, forearm and torso. For more information see e.g. Erp et al. (2005); for remote guidance (Aggravi et al. 2016), or in integration with sensor network technologies (RFID, Bluetooth and Wi-Fi) (Loconsole et al. 2016). In parallel, the effect of different parameters modulating vibrotactile stimulation on skin receptors have been evaluated in several studies, see e.g. Dakopoulos and Bourbakis (2009), Ando (2015), Scheggi et al. (2014).

Regarding spatial information, it can be gathered using various sensing modalities (Pissaloux et al. 2017; Gay et al. 2018), which include sonars (Villamizar et al. 2013), laser range finders or stereo cameras (Pundlik et al. 2011). Despite the promising results, the effective translation of these technologies in real-world applications is still limited. One of the causes for this scarce acceptance is related to the limited involvement of real users in the conceptual and design phases. Notwithstanding such tendency is changing, we are still far from systems designed for, and hence usable by, real people with real needs.

In this work, we propose a user-centered approach for the development of wearable technological solutions for indoor navigation, which moved from a preliminary investigation of the requirements of visually impaired people. A tight interaction with real end-users informed all the developmental phases of our system, driving us to the definition of the system layout, the cues used for delivering spatial-related information through sensory substitution, the implementation choices for the planning and navigation parts.

The wearable travel aid system for indoor navigation proposed in this work consists of: (1) a RGB-D camera that acquires both color and dense depth images from the environment; (2) a processing unit to process the images and perform the needed computations for obstacle avoidance and (3) a wearable fabric-based device, which can convey navigational information at the arm level to avoid the detected obstacles, through normal and tangential force delivery on the user's arm. This device represents a new version of the cutaneous passive haptic interface described in Barontini et al. (2024), which was specifically re-designed for this work to be more compact and light, and hence to meet users' requirements. The choice of this tactile device was motivated by the fact that the stimuli (i.e. pressure and skin-stretch) it can deliver are similar to the ones that a blind person would experience, if a volunteer would hold the person's arm to guide her/him through the environment.

Two different experiments were performed to validate the effectiveness of the presented navigation system: Experiments A (A1 and A2) with blindfolded participants, and Experiments B (B1 and B2) with blind participants. The results showed that our system can be a viable support for navigation, especially for subjects who are not expert users of the white cane.

5.2 The Navigation System

A questionnaire was conducted among blind individuals to gather insights on their orientation strategies, the desired features of a navigation device, and the types of information they needed immediately. The questions aimed to understand how blind people navigate unfamiliar environments, what functionalities they require in a navigation aid, and their preferences for receiving navigation cues. The key findings from the responses (see also Chap. 2) are summarized below:

- **Orientation strategies**: Participants indicated a preference for segmented and linear routes, which aid in better orientation and mobility. They emphasized the importance of creating mental maps by touching objects to avoid obstacles in unknown environments.
- **Device features**: Respondents highlighted the necessity of a wearable and hands-free system that would not hinder the use of a white cane or guide dog. Such a system should reliably detect all types of obstacles, including high-up obstacles and stairs, and communicate this information clearly to the user.
- **Type of information**: Haptic feedback was overwhelmingly preferred over auditory communication. Participants expressed the need to keep the auditory channel free to perceive environmental sounds, especially when navigating outdoor environments. Clear, simple, and instantly recognizable haptic signals were favored for conveying critical navigation information, such as changes in direction or obstacle detection.

The interface represents a fundamental component to manage the interaction between the system and the person. A good and appropriate interaction is one of the major goals of the Human-Computer Interaction (HCI) field, targeting the enhancement of accessibility and usability of a system or application.

Therefore, the design of the user interface and interaction features plays a crucial role in system development life cycle. For the navigation system presented in this work, we decided to rely on tactile stimulation—in agreement with the outcomes of the interviews—provided by a wearable system, to leave the hands free. In our case, the interface will hence consist of a wearable haptic device to deliver instructions and information to the user. As emerged from the interviews, important commands and instructions must be given in a clear manner. The translation of these requirements in terms of interface specifications can be summarized as follows:

- Haptic device, which is required to be comfortable and practical;
- Commands and instructions related to *stop*, *proceed*, *turn left/right*;
- Wearable components, which should be easy to use.

In the following, we will describe the key ingredients that we used for the hardware architecture of the travel aid for indoor navigation, i.e., the wearable haptic device and the sensing and processing unit.

5.2.1 The User Interface

The haptic interface on which we focused our attention based on the previously reported requirements is an engineered version of the clenching upper-limb force feedback wearable device for distributed mechano-tactile stimulation of normal and tangential skin forces (CUFF) described in Barontini et al. (2024). Briefly, the CUFF consists of two DC motors attached to a band or cuff worn around the user's arm. When the motors spin in opposite directions, they tighten or loosen the band on the arm, thus conveying a normal force. On the contrary, when the motors spin in the same direction, the fabric can slide around the arm, thus conveying tangential force cues and hence directional information, see Fig. 5.1a.

The usage of this type of system for navigation stimulus delivery was motivated by the idea of replicating the stimulation that the hand of an accompanying person provides to the blind person, to inform him/her about the presence of obstacles and on how to avoid them (and eventually to communicate a stop command). This information mainly relies on skin stretch, squeezing and tangential elicitation of the arm, see Fig. 5.1.

5.2.2 Vision Sensor and Processing Unit

The sensing apparatus of our integrated system mainly leverages the images acquired through a RGB-D camera placed on the user's chest, while a processing unit is placed on the back to process the visual information from the camera and provide commands to the CUFF, based on the obstacle avoidance algorithm described in Sect. 5.3, see Fig. 5.2. It is worth noting that, for this prototype, a light laptop was used as a cheap processing unit. In future developments, an ad–hoc unit will be used

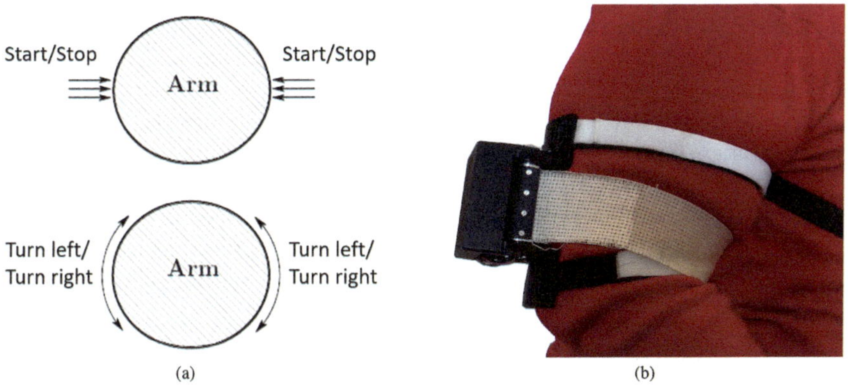

Fig. 5.1 Working modes of the CUFF (**a**) and overview of the device worn on the subject's arm (**b**)

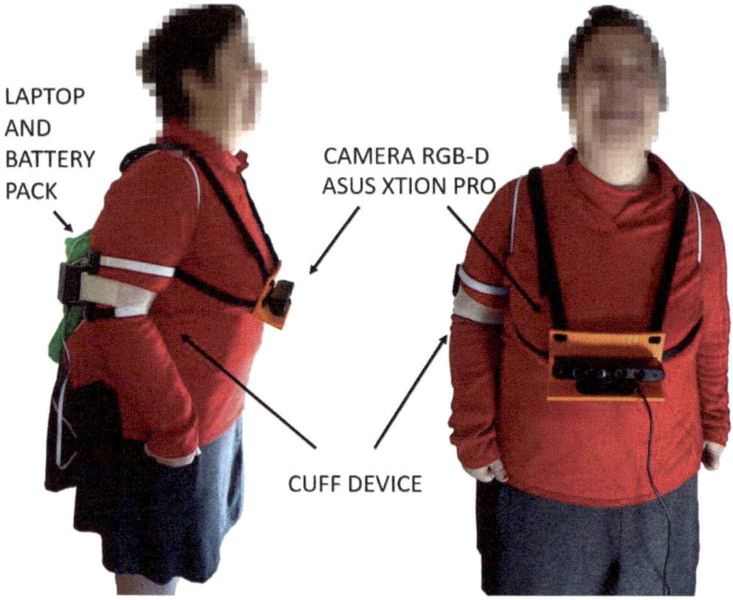

Fig. 5.2 The proposed navigation system worn by the user. We can see the wearable technological device, the camera and the processing unit (laptop): front view on the right, lateral view on the left

to further improve wearability. According to our original idea, the overall system should have included two cameras (to enable a more exhaustive space scanning) and two ultrasonic sensors per leg (placed at the calf level), to detect ground obstacles. Capitalizing upon the interview outcomes described in Chap. 2, we decided to modify the envisioned final layout of the system, to include the CUFF, only one RGB-D camera, placed on the user's body with a VELCRO belt at the chest level (in order to detect both high and medium-low obstacles), and a laptop to process the images, placed inside a backpack. The usage of additional cameras and sensors on the calf was avoided to reduce the complexity of the system and likely making the assistance more intuitive and comfortable.

We chose to use the *Asus Xtion Pro* camera, with a frame rate of 30 Hz at 640 × 480. However, after some preliminary testing, we realized that in this manner the stimulation provided by the CUFF was delivered at a frequency rate which could have generated confusion in participants. For this reason, we decided to halve the frame rate to 15 Hz, which is still sufficient for enabling a correct image processing. The communication between the different elements was implemented via USB connection.

5.3 The Obstacle Avoidance Strategy

To enable navigation in the environment, i.e. to reach a target position while simultaneously avoiding obstacles, we implemented an obstacle avoidance algorithm running on the laptop (ASUS UX310U). This algorithm can detect floating obstacles, such as people walking around, and enable the user to move in a safe manner, based on RGB-D data processing. The collision detection algorithm is based on Bach-y-Rita and Kercel (2003), where obstacles were found in images from a camera through corner detection, while the distance was provided by the depth sensor. We chose to implement the techniques in Bach-y-Rita and Kercel (2003) since they were proven to be both efficient and robust in detecting corners and suggesting a safe path to avoid the obstacles in experiments with blindfolded and blind people for indoor navigation. Based on the interview outcomes the navigation strategy was suitably modified as described in the following.

The detection algorithm is based on the subdivision of the captured image in three sub areas and a sequential obstacle detection in each area. Once an area is found to be obstacle free, the user is steered toward such area. Based on the interviews outcomes, we decided to customize the geometrical segmentation of the areas of interest. Indeed, since end users usually prefer to move straight, and to minimize the obstacle detection time, we identified a bigger central trapezoidal area, hereinafter named as *Safe Area*, and two lateral ones named *Right and Left Areas*, respectively. First, we looked for corners only in the Safe Area preventing the user to receive information on possible lateral obstacles that are negligible for safe navigation purposes. If an obstacle is detected in this area, the algorithm analyzed the corner in the Right and Left area sequentially. An example of the analysis of the corners inside the three subareas of investigation is shown in Fig. 5.3. In this way we limited the amount of information to the user coherently with the users' requirements we collected. The corner detection is

(a) (b)

Fig. 5.3 In these pictures it is possible to observe an example of the analysis of the corners inside the three subareas of investigation: central area (**a**), and lateral ones (**b**) based on the color information of the image, and it implements the OpenCV algorithm *Shi-Tomasi corner detector* reported in Shi (1994). The chosen *ASUS Xtion PRO* sensor provided point clouds with associated color and depth information for each point of the captured image.

Once corners have been extracted, the depth information of the selected points is used to determine the proximity of the potential obstacles with respect to the user. Based on the value extracted from the point cloud, we heuristically chose 3 different thresholds to detect close obstacle. It is worth noticing that since the white cane was selected as a needed tool by most of the interviewed persons, our choice was tailored on the condition that implies the usage of such assistive tool. However, this choice can be safely employed also in other cases of assisted autonomous navigation. The chosen thresholds are:

- If the depth values of all corners are greater than 2 m the corners represent obstacles whose distance from the user is sensibly greater than the length of the white cane. The area is hence classified as obstacle free (F).
- If at least one of the depth values is included in the range 1.2–2 m, the area is classified as occupied by an obstacle (O).
- If at least one of the depth values is less than 1.2 m, the area is classified as critical since there is a potential obstacle very close to the user (C).

When an obstacle is detected between the range 1.2–2 m the navigation algorithm starts to analyze the image, as described before, finding a free path and giving the command to turn to the right or to the left (R, L). Otherwise when an obstacle is detected closer than 1.2 m, the algorithm gives the STOP signal to the device. Then the user is commanded to rotate toward the area on the right, and if the path is not free, the commanded rotation provided by the CUFF is toward left.

Once a free path is found, a start command is given to re-start walking toward the goal direction—straight command (S). The described obstacle detection algorithm (ODA) takes as input one of the three areas (A) and returns as output the classification types of the analyzed area, i.e. free (F), occupied (O) or critical (C), and the commands to be implemented with the wearable devices, accordingly, i.e. straight (S), left (L), right (R), STOP, as described later. After a rotation command, if the path is free, the motors come back to the rest position. The navigation algorithm, based on the obstacle detection one, is shown in the Algorithm 1.

As previously mentioned, the algorithm outcomes are translated into tactile stimuli to be provided by the CUFF device on the user's right arm (see the accompanying video), as follows:

1. Straight walk (S) corresponding to two sequentially squeezing stimuli, which are implemented commanding opposite directions of rotation to the motors of approx 90°;
2. Turn left (L), corresponding to counterclockwise rotation of both motors of approx 180°;

Algorithm 1 Navigation Algorithm (NA)
Input: acquired image (I)
Output: straight (S), left (L), right (R), STOP, commands to be sent to the CUFF

```
if ODA(Safe Area)=F then
    send command S
    acquire new image I and execute NA(I)
else
    if ODA(Safe Area)=O then
        if ODA(Right Area)=F then
            send command R
            acquire new image I and execute NA(I)
        else
            if ODA(Right Area)=O then
                if ODA(Left Area)=F then
                    send command L
                    acquire new image I and execute NA(I)
                else
                    send STOP
                    acquire new image I and execute NA(I)
                end if
            else
                send STOP
                acquire new image I and execute NA(I)
            end if
        end if
    end if
    send STOP
    acquire new image I and execute NA(I)
end if
```

3. Turn right (R), corresponding to clockwise rotation of both motors of approx 180°;
4. Stop, corresponding single squeezing stimuli of higher intensity implemented with larger and opposite rotation of the motors of approx 360°.

These values were heuristically chosen as a good trade-off between intuitiveness and clarity of elicited perception and pleasantness. The CUFF device is able to elicit a maximum normal force of approx 25 N. For all the experiments, we chose for the Stop signal a maximum normal force of approx 18 N, and for the Start signal a normal force of approx 10 N. Regarding the Left and Right stimuli, we decided to pre-tension the belt of the CUFF device to provide a normal force approx 3 N, focusing more on the sliding motion elicited on the user's skin. In this case the commanded motor rotation was approximately of approx 4800 ticks. One of the four aforementioned commands is provided to the user, at every algorithm step, i.e. at a frequency of 15 Hz. For example, when in the central mask an obstacle is detected and on the right there is a free path, the command sent to the device is to turn right; untill the subject didn't change the trajectory the device continued to give him the command to turn right. If the path is free, the CUFF does not provide any stimulation to the user's arm. In this way, we avoid a constant CUFF activation, which may result in an annoying stimulation for the users as it results from the outcomes of our interviews. Furthermore, in agreement with the observations reported in the previous Section, we decided to give stimuli which result in segmented trajectories. It is worth noting that we chose a safe range of 1.2 m because the operating distance of the camera goes from 0.8–3.5 m; from 1.2 to 0.8 m the algorithm continues to give a STOP signal. If, accidentally, the distance from an obstacle goes under 0.8 m the camera is not able to see anything but the user is close enough to perceive the presence of the obstacle with the cane or hands. Thresholds can obviously be customized based on users' physical characteristics such as age, cognitive abilities and anatomical structure.

5.4 Experimental Validation

We tested the effectiveness of our system with different participants and end-users, in different environments and navigation conditions. The main goal of these experiments was to verify if the usage of our technological solution could potentially enhance autonomous walking performance in everyday life. For these reasons, in light of the importance of traditional navigation supports—in particular the white cane—for blind people, which emerged from our interviews, we decided to test our system alone, and also in association with the cane, and assess users' performance and impressions in the different cases. To ensure the correct implementation of the validation phase, we first evaluated the safety of our approach with blindfolded participants, gradually increasing the complexity of the path to perform and finally enrolling blind subjects. None of the participants had any physical or mental limitation which could have affected the experimental outcomes. The experimental procedure was approved by the local Ethical Committee of Area Vasta Nord Ovest (CEAVNO) protocol $n°$ 9219.

5.4.1 Experiments A: Participants

The Experiments A were completed only by blindfolded participants. We performed two different sessions of experiments: a first session where we involved 6 right-handed blindfolded participants—2 female, mean age 27, hereinafter referred to as Experiment A1; and a second session performed by 10 blindfolded participants—4 female, mean age 26 years, one left-handed hereinafter referred to as Experiment A2.

5.4.2 Experiments A1: Setup and Procedure

Experiment A1 was composed of: a discrimination task to evaluate if the directional stimuli provided by the wearable haptic device were clearly interpreted for navigation purposes, and a walking task in an unknown environment, where participants were asked to wear and use our navigation system to walk along a corridor, avoid obstacles (both fixed and moving) and reach a target goal. In both tasks participants were blindfolded.

Discrimination Task: We asked participants to wear the CUFF on the right arm and to recognize the direction of 50 randomized stimuli (50% for each direction). These stimuli were obtained by rotating of approx 180° the CUFF motors in the same direction, clockwise and counter-clockwise to suggest right and left rotation, respectively. The order of rotation direction was randomized. Blindfolded participants comfortably seated wearing the CUFF device, with the forearm placed on a table. They wore headphones with pink noise to prevent the usage of any auditory cue generated by the rotation of CUFF motors.

We decided to focus our investigation only on the left and right rotational stimuli because, in our opinion, these two stimuli could have been more challenging to interpret with respect to a strong squeeze on the arm that indicates a stop. Indeed, the interpretation of these rotational stimulations requires a proper association to a change of direction and then a suitable processing in participants' body schema.

Walking task: During this task, each subject was equipped with the CUFF on the right arm, the RGB-D camera on the chest, a laptop in the backpack and a white cane as in Fig. 5.1. The experiment was carried out in the second floor of the School of Engineering of Pisa, see Fig. 5.4a. Two different travel aid modalities were considered:

1. walking along the corridor using the white cane together with the CUFF
2. walking along the corridor using the white cane only.

A moving obstacle, i.e. a person walking toward the subject, was presented always in the same positions for both modalities.

In Fig. 5.6 a snapshot sequences of the experiments are reported (see also the accompanying video provided as supplemental material), where the user is able to avoid the obstacle based on the input of our navigation system. Before the experiment, subjects underwent a training period of fifteen minutes to understand the 4 commands delivered by the device, and how the navigation system worked. During this period, participants walked along a different part of the second floor of the building with respect to the one used for the experiments, to avoid any learning effect.

After the training, the subject completed the path in both modalities. The order of task execution for the two modalities were randomized and counterbalanced across subjects. A single trial experiment was performed by every participant for each modality in order to avoid learning effects.

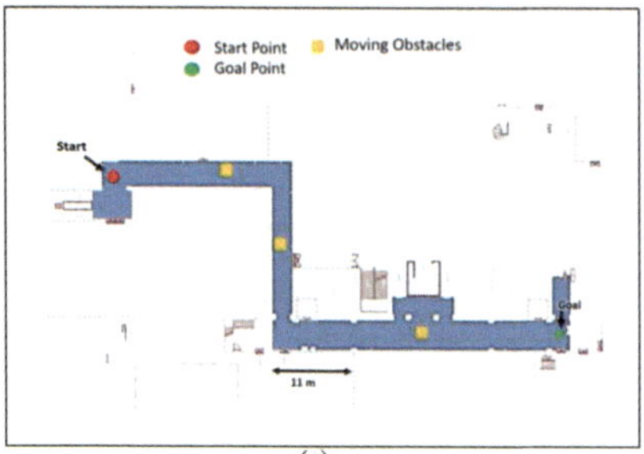

(a)

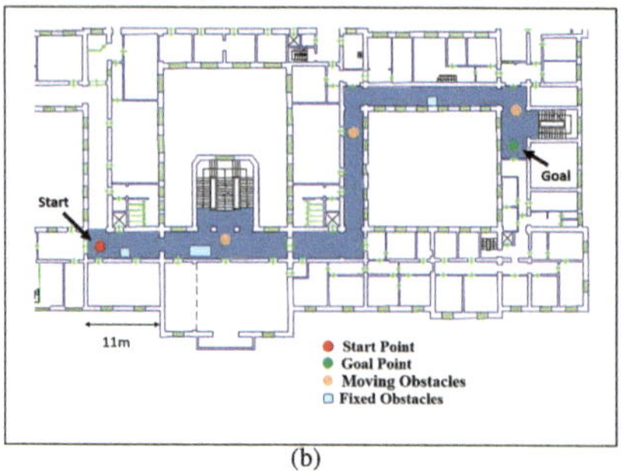

(b)

Fig. 5.4 **a** Map of the 2nd floor of School of Engineering of Pisa (Experiment A1, total path length approx 90 m). **b** Map of the 1st floor of School of Engineering of Pisa (Experiments A2, total path length approx 110 m). In blue, the path followed by participants during the experiments. The red and green points are the start and target position, respectively. The yellow points are the moving obstacles set for the modality experiments with the CANE only condition, the orange points are the moving obstacles set for the modality experiments with the CUFF plus CANE; the sky-blue points are the fixed obstacles present for each modality of the experiment. It is worth noting the moving obstacles were in the same locations for all the modalities

5.4.2.1 Experiment A2: Setup and Procedure

Experiment A2 consisted of a walking task only. It was conducted in the first and second floor of the School of Engineering in Pisa (see Fig. 5.4a, b) considering a longer and more complex path, with respect to the one used in Experiment A1, which also included a moving obstacle, i.e. a person walking towards the subjects. The test was divided in two parts:

1. training period to familiarize themselves with the system, as for Experiment A1;
2. walking period along the corridor in three different modalities (CUFF plus white cane, white cane only, CUFF only);

Ten blindfolded subjects wore our system as in Experiment A1. Also in this case a single experimental trial was performed for each condition. The moving obstacle was presented in the same points for all the modalities.

5.4.2.2 Experiments A: Data Collection

To evaluate the performance in Experiments A (completed by blindfolded participants), we recorded: the number of correct answers for the discrimination task (Experiment A1), the time for task accomplishment and the results of a subjective quantitative evaluation performed by administrating a Likert scale survey (Experiments A1 and A2).

Table 5.1 These statements, presented in random order, were rated by the blindfolded subjects of Experiments A (A1 and A2) using a 7-point Likert scale (1: Strongly disagree, 7: Strongly agree). Median and interquantile range across all individuals are reported

	Questions	Median	Iqr
Q1	It was easy to wear and use the CUFF device	6	2
Q2	It was easy to wear and use the CANE together with the cutaneous device	7	1
Q3	I was feeling uncomfortable while using the CANE together with the CUFF device	1.5	1
Q4	The haptic suggestions were intuitive	6	2
Q5	The sound produced by the actuators of the CUFF was hampering	2	1.5
Q6	The stimuli produced by the CUFF were easy to distinguish	6	2
Q7	The stimuli produced by the CUFF were helpful	7	1
Q8	It was easy to feel the presence of an obstacle receiving the feedback via the CUFF	6	1
Q9	I had the feeling of performing better while receiving feedback via the CANE only	1.5	1
Q10	I had the feeling of performing better while receiving feedback via the CANE and CUFF device	7	0.5
Q11	At the end of the experiment with the CUFF and the CANE I felt tired	1.5	1
Q12	At the end of the experiment with the CANE I felt tired	3.5	4

The time for task accomplishment is the time to reach target position as in Cardin et al. (2007), while the Likert scale consists of questions about the system and the experimental tasks, to which participants had to answer by assigning a score ranging from 1 *totally disagree* to 7 *totally agree*. This represents a common procedure to evaluate devices for assistive robotics and Human-Robot Interaction (**?**).

The questions delivered to the subjects are listed in Table 5.1, and the results are shown in Fig. 5.5. In particular, questions from Q1 to Q3 referred to the wearability of the CUFF device, questions from Q4 to Q8 investigated the intuitiveness of the sensory substitution, questions Q9 and Q10 were about the performance and questions Q11 and Q12 dealt with users' fatigue during the experiments.

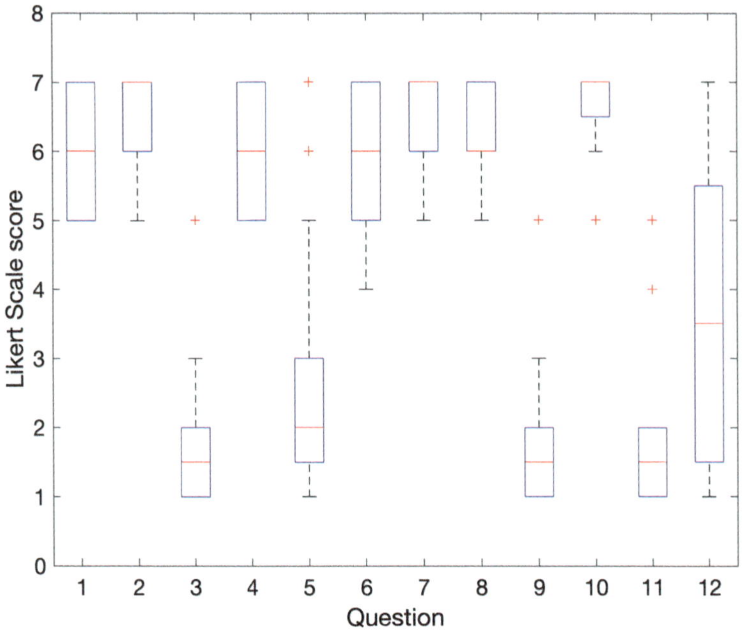

Fig. 5.5 The boxplot shows the results of the Likert Scale for the blindfolded participants. Center lines show the medians; box limits indicate the 25 and 75th percentiles; whiskers extend 1.5 times the interquartile range from the 25 and 75th percentiles. n = 12 blindfolded participants

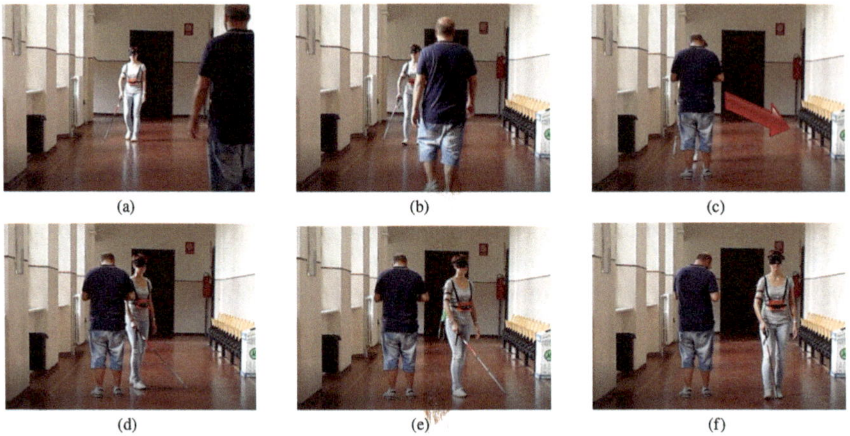

Fig. 5.6 In these pictures, it is possible to observe the obstacle avoidance, on the left, performed by the blindfolded individual while walking through the corridor

5.4.3 *Experiments B: Participants*

The Experiments B involved only blind participants and it consisted in two parts. The Experiment B1 included one blind subject, a 42-years-old female. Six blind subjects (3 female, mean age 51 years old, all congenitally blind) were involved in the Experiment B2. Four of them were expert users of the white cane; they navigated independently in outdoor and indoor unknown environments and only one of them preferred to use the white cane for outdoor navigation only.

The other participants do not use the white cane in their daily life, but a guiding volunteer. Two participants had previous experience with vibrotactile device for navigation and one of them used a vibrotactile feedback for outdoor navigation, i.e. the Miniguide (Sanz et al. 2018).

None of them had experience in the use of force feedback devices as sensory substitution techniques for travel aids. Participants gave their informed consent before the experiments, and none had any previous knowledge of the environment where the experiments were performed.

5.4.4 *Experiments B1: Setup and Procedure*

The Experiment B1 was completed by a single right-handed subject. The blind user performed the same experimental task and procedures as the blindfolded participants of Experiment A2 (for more information see Sect. 5.4.1 Experiment A2). The subject wore our system as all the other blindfolded participants. Since she does not use the white cane to move in everyday life but only a guiding volunteer, we did not consider the navigation condition with the cane only.

5.4.4.1 Experiment B2: Setup and Procedure

While, the Experiment B2 was completed by six blind subjects, all of them right-handed. They wore the CUFF on the right arm (only one user preferred to not have the device on the same side, right, of the cane), the RGB-D camera on the chest, a laptop in the backpack, and the white cane only for the expert users. The experiment was divided in three main parts:

- familiarization with the system and the tactile stimulation;
- experiment for recognizing the four stimuli (start, stop, turn left and turn right) provided via the CUFF;
- walking task.

Familiarization: Prior of each experiment, participants went through a training period with the system, of about three minutes, to take confidence with the device.

They were comfortably seated with the CUFF placed on their arm and the four commanded stimuli were provided sequentially in this order: start, turn right, stop, turn left.

The aim was only to give an idea of which kind of stimulus participants should have experienced via the CUFF, since none of them had previous experience with a force-haptic feedback device. After this part, participants spent 15 min walking along a corridor (first floor of the School of Engineering of the University of Pisa) to better understand the information provided by the CUFF, and the strategy adopted for obstacle avoidance.

The paths used for the familiarization were different from the ones used in the experimental tasks to avoid any learning effect.

Commands recognition task: In this task, we asked participants to identify the four commands delivered by the CUFF, while they were walking along a corridor. An operator commanded the CUFF device with a joystick, to deliver navigational cues and suggest the direction to follow. Participants had to recognize 40 randomized stimuli (ten for each of the four main commands).

After the analysis of the outcomes of the discrimination task in Experiment A1 (see Sect. 5.4.2), although the different stimuli were clearly recognized by participants, we received some comments on the usefulness of increasing the intensity of the stimulation for the navigation commands. For this reason, we decided to rotate the motors of approx 210° instead of approx 180° for the turn right and left commands, and about 360° for the stop and start conditions.

During the experiment we video-recorded the task execution to identify when the joystick buttons were pressed by the operator to command the CUFF device and obtain the reaction time of participants. The joystick used for this experiment was a common *PlayStation* joystick.

5.4.4.2 Navigation Task

To assess if our navigation system could be a viable solution not only in big and open spaces like the corridor used in the Experiments A and B1, we tested the system in five small walking tasks in narrow spaces, which included the navigation through a door, and multiple sequential changes of direction.

The aim of these tasks was to analyze: (1) the user's perception of the tactile commands, (2) the effectiveness of the system in detecting/avoiding the obstacles and (3) the effectiveness of the system in providing appropriate and precise instructions to guide the user. After a first pilot test, we decided to modulate the corridor of the second floor of the School of Engineering of Pisa with wood panels in order to create customized small paths, with various levels of navigation complexity:

- Task 1: to turn left
- Task 2 to turn right
- Task 3: to walk along a corridor and avoid a fixed obstacle

- Task 4: to change three times the direction (turn left, then right and finally turn left)
- Task 5: to pass through a door.

Tasks 1 and 2 targeted two different direction changes; Task 3 dealt with obstacle avoidance; Task 4 required multiple sequential changes of direction, and Task 5 was related to the navigation through a door. The paths were approx 11 m long (Tasks 1, 2, and 4), and approx 7 m (Tasks 3 and 5), while the width was approx 1.6 m for all the paths.

Since not all the participants were expert users of the white cane, we decided to take in consideration two different modalities of execution. We left the cane to the three expert users so they repeated every task twice: once using the CUFF in conjunction with the cane, and another using only the cane as a travel aid. On the contrary, the other three participants performed the tasks only with the CUFF. It is worth reporting that the expert users of the white cane in Experiment B2 were in total four, but one of them decided to not use the cane during the experimental session, to be completely focused on the CUFF device. The order of task execution for cane users was randomized and counterbalanced across subjects for the two different modalities.

5.4.4.3 Experiments B: Data Collection

For the Experiment B1, we recorded the time for task accomplishment and the results of a subjective quantitative evaluation performed by administrating a Likert scale survey (like in Experiments A1 and A2). Since the participant was not an expert user of the cane, the questions focused on the navigation condition with only the CUFF as a support tool, see Table 5.7 for more details. Furthermore, we asked the participant if she would have preferred the usage of auditory or vibratory cues with respect to forces as a sensory substitution method.

Regarding the Experiment B2, to evaluate users' navigation performance related to the clarity and the effectiveness of the commands provided via the device during the recognition task, we decided to record the number of correct and incorrect movements (based on the agreement with the delivered navigational stimulation) and also the reaction time to the stimuli. We defined the reaction time as the time interval between the instant when the joystick button was pressed by the operator and the reaction of the subject as it can be observed from the video (which used a frame rate of 15 Hz). For each navigation experiment (Tasks 1 to 5), we recorded the time for task accomplishment, and the number of collisions with the obstacle in case of failure of the vision system (please note that the camera supports a near field mode at least of 0.8 m). A single trial experiment was performed by subjects for each modality to avoid any learning effect.

At the end of the experiment participants underwent through a subjective quantitative evaluation procedure based on a seven point Likert scale, (see Table 5.1). We divided the questions for the blind subjects based on the modalities of task execution. Questions from Q1 to Q14 were in common to all the subjects. In particular, questions from Q1 to Q11 referred to the wearability of the CUFF, the intuitiveness of sensory substitution and the sensation provided by the device, whereas questions from Q12 to Q14 investigated participants' preferences with respect to other types of sensory substitution cues, like auditory stimuli or vibration. Questions from Q15 to Q18 were delivered to the participants who completed the walking task with the CUFF only, while questions from Q19 to Q28 were delivered to the participants who completed the walking task in both modalities.

5.5 Experimental Results

Starting from Experiments A, we can see that the results of the discrimination task had an average percentage of 97% and a standard deviation of 3.41. Table 5.1 reports the Likert scale results for the blindfolded subjects in the Experiments A1 and A2. The questions Q1 and Q2 (related to the wearability of the device with and without the use of the cane) were positively rated with a median of 6 and 7 and an interquantile range of 23 and 1 respectively. For Q6 and Q7, related to the usefulness of the stimuli, the median responses were 6 and 7 with a range of 2 and 1. Questions Q9 and Q10, on the performance, showed positive results with a median of 1.5 and 7 and a range of 1 and 0.5. For Q11 (i.e. the fatigue at the end of the experiment) the median was 1.5 and the range was 1.

The differences between the time for task accomplishment in the two navigation conditions were studied using a Wilcoxon signed-rank test, due to the non-Gaussianity of the samples. In particular, the p-value from Wilcoxon non-parametric test is associated with the null hypothesis of equal median values. Results show that there is not a significant difference in term of time for task accomplishment in the two navigation conditions ($p > 0.1$, Experiment A1: Time with Cane only seconds: mean standard 165 deviation 42.67; Time with CUFF and Cane: mean 162.5 standard deviation 27.5. Experiment A2: Time with Cane only seconds: mean 401 standard deviation 170.8; Time with CUFF and Cane: mean 343 standard deviation 67.34). Furthermore we performed a Wilcoxon non-parametric test considering scores from Q9–Q10, and Q11–Q12, respectively. Statistically significant differences were found in both cases ($p < 0.05$), which suggest that the integration with the CUFF was perceived as effective in improving the navigation performance and reducing the fatigue. Table 5.7 reports the result for the blind participant of Experiment B1, while Table 5.2 shows the time for task accomplishment for Experiments A and B1. The time for task completion of the blind subject in Experiment 1 was 180 sec.

Regarding the Experiment B2, Tables 5.3 and 5.4 show the time of reaction of the participants to the commands and the mean value across all subjects. As it can be noticed we report data from only five blind subjects; we decided to discard the sixth subject since we experienced some issues during the experimental task due to communication problems from the joystick to the CUFF. What is noticeable is that

Table 5.2 The table shows the time in seconds for accomplishing the walking task for Experiment A1 and for the Experiment A2 and B1 (only one blind subject participated in the latter experiment for the CUFF-only condition. The corresponding value is reported in red in the last row of the table)

	Experiment A1		Experiment A2 and B1	
	Mean	Std. Dev.	Mean	Std. Dev.
CUFF plus cane	162.5	27.5	343	67.34
Cane only	165	42.67	401	170.8
CUFF only	/	/	180	

Table 5.3 This table reports the reaction time for each subject and the mean value associated to each CUFF command

Commands	Subjects					Mean	Std. Dev.
	S1	S2	S3	S4	S5		
START	2.17	1.44	1.51	1.29	1.62	1.61	0.33
STOP	1.28	1.03	1.03	0.9	1.04	1.1	0.13
RIGHT	0.84	1.66	1.08	1.06	1.39	1.3	0.32
LEFT	0.78	0.67	0.94	1.39	0.99	0.87	0.14

Table 5.4 This table reports the number of stimuli not correctly understood by the subjects

Commands	Subjects				
	S1	S2	S3	S4	S5
START	0	0	0	0	0
STOP	0	0	1	0	1
RIGHT	0	0	0	1	1
LEFT	1	0	1	0	2

participants promptly responded to the navigation commands with a mean time of reaction of 1.61 s for start command, 1.1 for the stop, 1.3 for turn right and 0.87 for turn left. This result testifies in favor of the intuitiveness of the delivered stimuli.

Table 5.5 reports the execution times for Task 1 to 5 in the different navigation modalities, while Table 5.6 shows the number of times in which the vision system failed, which testifies in favor of the reliability of our solution. The mean time and standard deviation, for the blind participants who were not expert users are: 70, 64, 71.5, 73, 72.5 s and 52.32, 35.35, 31.81, 31.11, and 38.89 s respectively for Task 1, 2, 3, 4, and 5. The blind participants, who were expert users of the cane completed the tasks in two modalities (CUFF + Cane and Cane only). The mean values and standard deviation are: 45, 50, 24.66, 38.66 and 40 s and 24.33, 35.79, 6.5, 21.54, 27.49 s for the modality CUFF + Cane. For the Cane only they are 19.33, 16.66, 13.33 17.33 and 14 s for the mean values and 1.15, 2.3, 1.52, 2.51 and 5.56 s for the standard deviation values (Table 5.7).

Table 5.5 The table shows the time (in second), mean and standard deviation to accomplish the five navigation tasks for each different experimental modality

Commands	CUFF only		CUFF + Cane		Cane only	
	Mean	Std. Dev.	Mean	Std. Dev.	Mean	Std. Dev.
Task 1	70	52.32	45	24.33	19.33	1.15
Task 2	64	35.35	50	35.79	16.66	2.3
Task 3	71.5	31.81	24.66	6.5	13.33	1.52
Task 4	73	31.11	38.66	21.54	17.33	2.51
Task 5	72.5	38.89	40	27.49	14	5.56

Table 5.6 The table reports the number of collisions, mean and standard deviation in each task for the different modalities

Commands	CUFF only		CUFF + Cane		Global	
	Mean	Std. Dev.	Mean	Std. Dev.	Mean	Std. Dev.
Task 1	1	0	1.5	0.57	1.4	0.54
Task 2	0.5	0.7	0.33	0.57	0.4	0.54
Task 3	0.5	0.7	0.66	6.5	0.6	0.54
Task 4	2.5	0.7	1.33	0.57	1.8	0.83
Task 5	2	0	0.66	1.15	1.2	1.09

Table 5.7 These statements, presented in random order, were rated by the blind subject of Experiment B1 using a 7-point Likert scale (1: Strongly disagree, 7: Strongly agree)

	Questions	Results
Q1	It was easy to wear and use the CUFF device	7
Q2	I was feeling uncomfortable while using the CUFF device	2
Q3	The haptic suggestions were intuitive	7
Q4	The sound produced by the actuators of the CUFF was hampering	1
Q5	The stimuli produced by the CUFF were easy to distinguish	5
Q6	The stimuli produced by the CUFF were helpful	7
Q7	It was easy to feel the presence of an obstacle receiving the feedback by the CUFF	6
Q11	At the end of the experiment with the CUFF I felt tired	1
Q12	Have you ever used a vibrotactile ETA? If yes, you would have preferred it w.r.t the CUFF	Yes; No
Q13	Would you have preferred an auditory cue?	No
Q14	Would you use our system in everyday life?	Yes

Table 5.8 These statements were rated by the subjects of Experiment B2 using a 7-point Likert scale (1: Strongly disagree, 7: Strongly agree). Median and interquantile range across all individuals are reported

	Questions	All participants		CUFF only users		Cane users	
		Median	Iqr	Median	Iqr	Median	Iqr
Q1	It was easy to wear and use the CUFF device	6.5	3	/	/	/	/
Q2	The haptic suggestions were intuitive	4	2	/	/	/	/
Q3	I felt hampered by the cutaneous device	2	4	/	/	/	/
Q4	The sound produced by the actuators of the CUFF was hampering	1	0	/	/	/	/
Q5	The stimuli produced by the CUFF were easy to distinguish	5	1	/	/	/	/
Q6	The stimuli produced by the CUFF were helpful	3	4	/	/	/	/
Q7	I felt confident using the system	4	3	/	/	/	/
Q8	I felt stressed/frustrated using the system	2	3	/	/	/	/
Q9	The system gave me an unpleasant feeling	1	3	/	/	/	/
Q10	I was able to interact with the environment during the use of CUFF	6	2	/	/	/	/
Q11	I had difficult in locating the obstacles during the task	4.5	2	/	/	/	/
Q12	Would you prefer an auditory stimulus?	2 yes only	6 no	/	/	/	/
Q13	Do you ever use a vibrotactile device? If yes would you have preferred a vibrotactile stimulus?	2 yes	6 no	/	/	/	/
Q14	Do you wear the system during the daily life?	4 yes	4 no	/	/	/	/
Q15	I felt relaxed using only the CUFF device during the tasks.	/	/	7	1.5	/	/
Q16	I was feeling uncomfortable while using the CUFF device.	/	/	2	3	/	/
Q17	At the end of experiment with the CUFF only I felt tired.	/	/	4	3.75	/	/
Q18	I had difficult in completing the tasks using the CUFF only	/	/	5	2.25	/	/

(continued)

Table 5.8 (continued)

	Questions	All participants		CUFF only users		Cane users	
		Median	Iqr	Median	Iqr	Median	Iqr
Q19	It was easy to wear and use the CANE together with the CUFF.	/	/	/	/	7	0.75
Q20	I was feeling uncomfortable while using the CANE together with the CUFF device	/	/	/	/	1	0.75
Q21	At the end of experiment with the CUFF and CANE I felt tired.	/	/	/	/	1	3.75
Q22	At the end of experiment with the CANE only I felt tired.	/	/	/	/	1	0
Q23	I had the feeling of performing better while receiving feedback by the CANE only.	/	/	/	/	6	1.5
Q24	I had feeling of performing better while receiving feedback by the CANE and CUFF device.	/	/	/	/	4	2.25
Q25	I had difficult in completing the tasks using the CANE with the CUFF.	/	/	/	/	4	3
Q26	I had difficult in completing the tasks using the CANE only.	/	/	/	/	1	0
Q27	Using the CUFF with the CANE I was more hampered in completing the task	/	/	/	/	6	4.5

In Table 5.8 and in Fig. 5.7, we can find the results of the Likert scale questionnaire provided by all the blind participants of Experiment B2. Questions Q1, Q19 and Q20, are related to the wearability of the CUFF, where we can observe positive scores, with a median value of 6.5, 7 and 1, and a range of 3, 0.75 and 0.75, respectively. For Q5 and Q6 (on the sensory substitution) the median is 5 and 3, and the standard deviation 1 and 4. Questions Q15 and Q16, on the sensation during the usage of the device, also comes with a good rate with a median of 7 and 2, and a range of 1.5 and 3, respectively. For Q25 and Q26, on the task performance, the mean are 4 and 1, and the interquantile range of 3 and 0.

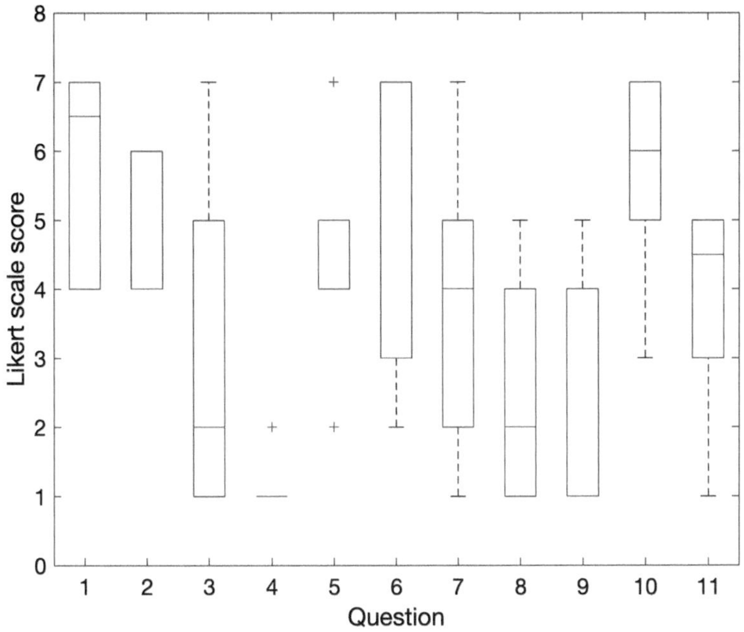

Fig. 5.7 The boxplot shows the results of the Likert Scale for the blind participants. Center lines show the medians; box limits indicate the 25 and 75th percentiles; whiskers extend 1.5 times the interquartile range from the 25 and 75th percentiles. n = 5 blind participants

5.6 Discussion

In this section, we discuss the findings from our experimental validation, divided into two main parts. The first part focuses on the results obtained from experiments involving healthy subjects, while the second part examines the outcomes from experiments conducted with blind subjects. This division allows for a clear understanding of how the system performs across different user groups.

5.6.1 Results with Blindfolded Participants

The results of the discrimination task of Experiment A1 showed that participants were able to correctly recognize the direction of the stimuli. Regarding Table 5.1 we can note that participants had no difficulty wearing and using the CUFF, also in combination with the cane. About the intuitiveness of the proposed sensory substitution approach, users were able to distinguish the stimuli produced by the device and understand the correct direction to follow. The general opinion was that haptic stimuli were helpful for guidance, especially for obstacle avoidance. All the subjects were able to avoid obstacles by following the information provided by the device even though some subjects reported that the sound produced by the actuator of the CUFF was hampering. Since the participants were not expert users of the cane, they had the impression of performing better when they used the cane and the CUFF together, instead of the cane only. Users did not feel tired at the end of experiments with the CUFF and the cane, while on the contrary they had some difficulties to accomplish the experiment with the cane only.

In conclusion, the integrated usage of the CUFF and the white cane did not degrade the performance in navigation task accomplishment with respect to the white cane only, but, on the contrary, increases users' confidence in performing in a better manner, especially for obstacle avoidance, as reported in the subjective quantitative evaluation

5.6.2 *Result with Blind Participants*

The results of the blind subject in Experiment B1 are reported in Table 5.7. The subject perceived the CUFF as highly wearable and easy to use (Q1), comfortable (Q2), intuitive (Q3), helpful (Q6, Q7), and at the end of the experiment the participant did not feel tired at all (Q11). The subject reported that the system could be used in her everyday life but wearability should be improved.

It is important to observe that the blind participant performed the experiment with the CUFF modality only since she was not an expert user of the cane. She confirmed that with the CUFF she felt safer walking through the corridor (Q7); about the sensory substitution, she declared that the haptic stimuli were easy to distinguish and helpful and intuitive (questions from Q3 to Q6).

The results of the experiments conducted with the blind participants, reported in Tables 5.3 and 5.6, showed that the commands were easy and clear to distinguish. What emerged at the end of the task is that the motor movement back to the reference position can be sometimes misunderstood as an indication to turn left or right. A possible solution could be the usage of two CUFF, one for each arm; or a different control strategy to move back to the reference position the CUFF's motors. It is reasonable to think that this could make users stop, or just slow down their pace. What we learnt from the experiments was that the participants who are expert users of the cane tended to slow down their pace to be more focused on the stimuli provided by the device. This can be interpreted as a natural adaptation to the new assistive modality, with respect to the one used in everyday life. On the contrary, the non expert users of the white cane walked with their normal pace. This observation testified in favor of the intuitiveness of the sensory substitution we chose. For what concerns the walking task, results reported in Table 5.5 show that all the subjects were able to complete the tasks with differences in terms of time for task accomplishment; however, due to the reduced number of subjects a statistical analysis cannot be performed. Expert users of the white cane were able to find the free path immediately using only the cane, while they had to spent a few seconds more to understand the indication received from the CUFF and pay attention to the command to follow. On the other hand, participants, who used a volunteer guide in everyday life, were able to complete the task slowly but autonomously.

Regarding Table 3.5, we can note that in general participants had no difficulty to wear and use the device, neither in conjunction with the cane (Q1, Q19 and Q20). About the intuitiveness of the sensory substitution, the stimuli were helpful and easy to distinguish for the participant who rely on an accompanying person in everyday life, less intuitive and useful for the expert users of the cane (Q5 and Q6); but all of them were able to successfully localize and avoid the obstacles during the tasks. More specifically, the global impression of users of white cane was that the device could be a viable solution for training new blind people in the usage of the cane, but to expert users of the cane the device seems to not give any additional information (Q23 and Q24). Indeed, they liked very much the idea of receiving information about the presence of obstacles, but they preferred to choose the best strategy to avoid it by themselves using the cane.

About the non autonomous blind participants, who rely on an accompanying person, all of them liked the device because for the first time they walked alone in an unknown environments. They felt relaxed using the system and understood well the stimuli (Q15 and Q16); although they declared to have some difficulties to accomplish challenging tasks, like passing through the door and double turning (see Table 5.6). For this reason at the end of these experiments they felt tired (Q17).

Furthermore, we asked the visually impaired users if they would have preferred forces versus vibratory cues: results show that they did not like this type of stimulus, and preferred to have the auditory channel free. Indeed, as said in Sect. 5.4.3, some of them had, in other occasions, tried a vibration-based device. Even though most of them had preferred the forces cues, in their opinion the most important thing is related to the fact that the stimulus has to be clearly understood without any possibility of misunderstanding.

5.7 Limitation and Future Developments

This work reports on the feasibility of the integration of a wearable haptic device, which provides stimulation cues similar to the ones conveyed by the hand of an accompanying person, with an obstacle avoidance systems. The work has followed a user-centered approach, from the definition of the specifications to the validation phase.

Despite the positive results, we are aware that the process for translating these outcomes in everyday life is still long and that our approach presents some limitations, which we would like to address in future work. The main aspects we would like to further develop are: (1) the user's requirements—social acceptability of the system, which is also related to its wearability, (2) the hardware architecture, and (3) the algorithm used for the obstacle detection. Regarding the first point, we will explore the possibility of tailoring the navigation instructions on different conditions of the end users, which are based on the travel support used in their daily activities. Indeed, what we learned from the experimental outcomes is that different guidance instructions could be needed for people using the white cane and for those who do not. Furthermore, what emerged from the outcomes of the Likert Scale (see question Q14 in Table 5.7) is that, even if the users appreciate our approach, they would like to have a system with smaller physical dimensions and more visually unnoticeable for an effective usage in their everyday life. This aspect, which is intertwined with the wearability of our architecture, was also highlighted during the informal discussions we had with the participants. This point could be achieved by exploring the exploitation of a processing unit that can be more easily integrated with the user's body, and of other types of sensors with a reduced layout, e.g. depth, proximity sensors. All these investigations and possible changes will be driven by a continuous feedback from visually impaired people. A more in depth analysis of the sensing solutions could represent a strategy to overcome the limitations of the current hardware (point 2), in particular the reduced capacity of the camera to generalize and see in different scene illumination conditions. Indeed, the camera and consequently the algorithm used in our system result in robust in environments that are illuminated with artificial light. On the contrary, in natural light settings, the reflection of the sunlight on the floor could lead to the misleading identification of false obstacles. Moreover, the camera is completely blind in a dark environment. Regarding point (3), the major limit of our algorithm is that it is not capable to distinguish between moving or fixed obstacles. This could not guarantee a safe navigation in a crowded place. Also this point could be addressed through a suitable choice of the sensing system. Under this regard, we will evaluate the integration and usage of vision-based and non-vision based sensors and arrays. This evaluation will include the investigation of a multi-sensory approach to increase the robustness of on-line obstacle identification and localization and people detection, for an intuitive spatial map reconstruction (Wang et al. 2017; Wachaja et al. 2015; Carballo and Ohya 2008; Cosgun et al. 2014). Although multi-sensory integration will be investigated, a minimalistic approach for resource usage will always guide all the phases, to push further system wearability and cost-effectiveness. Furthermore to increase the reliability of the algorithm, we will investigate a new lower level planning strategy based on a learning process on the pace of the users, and create a custom lower level planner. This could also help to increase system acceptability. An integration between the navigation and planning strategy presented in this document with a higher level planning method for the automatic computation of intermediate target points to drive the user is also envisioned. This could be done through the realization of a database with maps of different places of interest (e.g. hospitals, public offices, etc.). The testing of other navigation algorithms coming with a lower computational cost is under investigation, and applications for outdoor navigation will be considered. Finally, future work will include additional experiments with blind people and further investigation and implementation of users' needs, to continuously increase the intuitiveness of the system and its portability, including specific experiments to evaluate the effectiveness of the system for training blind people to use travel aids (such as the white cane).

5.8 Conclusion

In this work, we have presented a wearable navigation system of a RGB-D camera, a laptop and a wearable device CUFF, which can provide normal and tangential force cues through the control of a stretching state of a fabric. The target application is blind people guidance in an unknown indoor environment. The final layout of the system and obstacle avoidance techniques was based on the elaboration of requirements and opinions collected from blind individuals and people working in the field of assistance of the visually impaired. We tested the system with blindfolded participants and blind users, in different indoor environments, and verified if it could be a viable solution to increase performance of users with regard to autonomous navigation with and without the white cane usage (see Experiments A, B). Experiments show that our navigation system could be a viable solution to be integrated with classic navigation methods. Interestingly, the visually impaired people, which performed the experiment with the CUFF only (e.g. no cane users) exhibited a good time for task accomplishment and a positive perception of the navigation system and haptic stimuli. On the other hand, according to the blind expert users of the white cane, the device could be a valid aid for training newly blind people to use the white cane. Furthermore the users well recognized the stimuli provided by the CUFF, and they considered them helpful for navigation.

Acknowledgements The authors would like to thank the association "Unione Italiana dei Ciechi ed Ipovedenti ONLUS (UIC)" for the availability and help in recruiting blind participants for this study. A special thank goes to the visually impaired people who spent their time to test our system. Finally, they would also like to thank Prof. Daniela Rus (MIT), for the inspiring discussions.

References

Adame MR, Yu J, Moller K, Seemann E (2013) A wearable navigation aid for blind people using a vibrotactile information transfer system. In: 2013 ICME international conference on complex medical engineering. IEEE, pp 13–18

Aggravi M, Salvietti G, Prattichizzo D (2016) Haptic assistive bracelets for blind skier guidance. In: Proceedings of the 7th augmented human international conference 2016, pp 1–4

Andò B, Baglio S, Marletta V, Valastro A (2015) A haptic solution to assist visually impaired in mobility tasks. IEEE Trans Human-Mach Syst 45(5): 641–646

Bach-y-Rita P, Kercel SW (2003) Sensory substitution and the human-machine interface. Trends Cogn Sci 7(12):541–546

Barontini F, Catalano MG, Fani S, Grioli G, Bianchi M, Bicchi A (2024) The cuff, clenching upper-limb force feedback wearable device: design, characterization and validation. IEEE Trans Haptics

Borenstein J, Ulrich I (1997) The guidecane-a computerized travel aid for the active guidance of blind pedestrians. In: Proceedings of international conference on robotics and automation, vol 2, pp 1283–1288. IEEE

Carballo A, Ohya A (2008) Fusion of double layered multiple laser range finders for people detection from a mobile robot. In: 2008 IEEE international conference on multisensor fusion and integration for intelligent systems. IEEE, pp 677–682

Cardin S, Thalmann D, Vexo F (2007) A wearable system for mobility improvement of visually impaired people. Visual Comput 23:109–118

Cosgun A, Sisbot EA, Christensen HI (2014) Guidance for human navigation using a vibro-tactile belt interface and robot-like motion planning. In: 2014 IEEE international conference on robotics and automation (ICRA). IEEE, pp 6350–6355

Dakopoulos D, Bourbakis NG (2009) Wearable obstacle avoidance electronic travel aids for blind: a survey. IEEE Trans Syst Man Cybern Part C (Applications and Reviews) 40(1):25–35

Della Santina C, Piazza C, Gasparri GM, Bonilla M, Catalano MG, Grioli G, Garabini M, Bicchi A (2017) The quest for natural machine motion: an open platform to fast-prototyping articulated soft robots. IEEE Robot Autom Mag 24(1):48–56

Erp JBV, Veen HAV, Jansen C, Dobbins T (2005) Waypoint navigation with a vibrotactile waist belt. ACM Trans Appl Perception (TAP) 2(2):106–117

European Forum Against Blindness, EFAB (2013) [Online]. Available: https://www.euractiv.com/section/health-consumers/news/single-eu-definition-of-blindness-could-help-save-billions-of-euros-study/

Gay S, Rivière MA, Pissaloux E (2018) Towards haptic surface devices with force feedback for visually impaired people. In: International conference on computers helping people with special needs. Springer International Publishing, Cham, pp 258–266

Kassim AM, Yasuno T, Suzuki H, Jaafar HI, Aras MSM (2016) Indoor navigation system based on passive RFID transponder with digital compass for visually impaired people. Int J Adv Comput Sci Appl 7(2)

Keyes A, D'Souza M, Postula A (2015) Navigation for the blind using a wireless sensor haptic glove. In: 2015 4th mediterranean conference on embedded computing (MECO). IEEE, pp 361–364

Loconsole C, Dehkordi MB, Sotgiu E, Fontana M, Bergamasco M, Frisoli A (2016) An IMU and RFID-based navigation system providing vibrotactile feedback for visually impaired people. In: Haptics: perception, devices, control, and applications: 10th international conference, EuroHaptics 2016, London, UK, Proceedings, Part I 10. Springer International Publishing, pp 360–370

Meshram VV, Patil K, Meshram VA, Shu FC (2019) An astute assistive device for mobility and object recognition for visually impaired people. IEEE Trans Human-Mach Syst 49(5):449–460

Moller K, Toth F, Wang L, Moller J, Arras KO, Bach M, Schumann S, Guttmann J (2009) Enhanced perception for visually impaired people. In: 2009 3rd international conference on bioinformatics and biomedical engineering. IEEE, pp 1–4

Pacchierotti C, Sinclair S, Solazzi M, Frisoli A, Hayward V, Prattichizzo D (2017) Wearable haptic systems for the fingertip and the hand: taxonomy, review, and perspectives. IEEE Trans Haptics 10(4):580–600

Pissaloux EE, Velázquez R, Maingreaud F (2017) A new framework for cognitive mobility of visually impaired users in using tactile device. IEEE Trans Human-Mach Syst 47(6):1040–1051

Pundlik S, Peli E, Luo G (2011) Time to collision and collision risk estimation from local scale and motion. In: Advances in visual computing: 7th international symposium, ISVC 2011, Las Vegas, NV, USA, September 26–28 (2011) Proceedings, Part I 7. Springer, Berlin, pp 728–737

Rodríguez JL, Velázquez R, Del-Valle-Soto C, Gutiérrez S, Varona J, Enríquez-Zarate J (2019) Active and passive haptic perception of shape: passive haptics can support navigation. Electronics 8(3):355

Saegusa S, Yasuda Y, Uratani Y, Tanaka E, Makino T, Chang JY (2011) Development of a guide-dog robot: human-robot interface considering walking conditions for a visually handicapped person. Microsyst Technol 17:1169–1174

Sanz PR, Mezcua BR, Pena JMS (2018) ICTs for orientation and mobility for blind people: a state of the art. In: Wearable technologies: concepts, methodologies, tools, and applications, pp 1177–1203

Scheggi S, Talarico A, Prattichizzo D (2014) A remote guidance system for blind and visually impaired people via vibrotactile haptic feedback. In: 22nd mediterranean conference on control and automation. IEEE, pp 20–23
Shi J (1994) Good features to track. In: 1994 Proceedings of IEEE conference on computer vision and pattern recognition. IEEE, pp 593–600
Smith CM (1997) Human factors in haptic interfaces. XRDS: Crossroads, ACM Mag Stud 3(3):14–16
SungMin H (2016) Haptic aids for the visually impaired. In: Pervasive haptics: science, design, and application, pp 221–229
Velázquez R (2010) Wearable assistive devices for the blind. In: Wearable and autonomous biomedical devices and systems for smart environment: issues and characterization, pp 331–349
Villamizar LH, Gualdron M, Gonzalez F, Aceros J, Rizzo-Sierra CV (2013) A necklace sonar with adjustable scope range for assisting the visually impaired. In: 2013 35th annual international conference of the IEEE engineering in medicine and biology society (EMBC). IEEE, pp 1450–1453
Wachaja A, Agarwal P, Zink M, Adame MR, Möller K, Burgard W (2015) Navigating blind people with a smart walker. In: 2015 IEEE/RSJ international conference on intelligent robots and systems (IROS). IEEE, pp 6014–6019
Wahab MHA, Talib AA, Kadir HA, Johari A, Noraziah A, Sidek RM, Mutalib AA (2011) Smart cane: assistive cane for visually-impaired people. arXiv:1110.5156
Wang HC, Katzschmann RK, Teng S, Araki B, Giarré L, Rus D (2017) Enabling independent navigation for visually impaired people through a wearable vision-based feedback system. In: 2017 IEEE international conference on robotics and automation (ICRA). IEEE, pp 6533–6540
Wang Y, Kuchenbecker KJ (2012) HALO: Haptic alerts for low-hanging obstacles in white cane navigation. In: 2012 IEEE haptics symposium (HAPTICS). IEEE, pp 527–532
Wei Y, Kou X, Lee MC (2014) A new vision and navigation research for a guide-dog robot system in urban system. In: 2014 IEEE/ASME international conference on advanced intelligent mechatronics. IEEE, pp 1290–1295
World Health Organization, WHO (2014) Visual impairment and blindeness—Fact Sheet N 282, [Online]. Available: http://www.who.int/mediacentre/factsheets/fs282/en. Accessed 30 Nov 2016
Zelek JS (2005) Seeing by touch (haptics) for wayfinding. In: International congress series, vol 1282, pp 1108–1112. Elsevier

Chapter 6
Can Force Feedback Improve the Control of a Neuroprosthesis?

Abstract In this work, we present the Clenching Upper-Limb Force Feedback Device (CUFF), integrated with the SoftHand Pro robotic hand. Grasping an object is a complex action that relies on sensory feedback. While prosthetic hands assist with mechanical performance, they often lack adequate sensory feedback. Providing feedback about a prosthetic hand's grasp force is crucial for individuals with limb loss. We tested the CUFF system, controlled via myoelectrics of the forearm muscles, with five participants with limb loss and nineteen able-bodied participants. They completed a constrained grasping task (with and without feedback) requiring modulation of the grasp to reach a target force, with vision and hearing significantly limited. Data were analyzed using generalized linear models and Functional Principal Component Analysis (fPCA). CUFF feedback improved the able-bodied participants' ability to achieve the target grasp force and enhanced grasp precision for participants with limb loss who typically use body-powered prostheses, as well as for a subset of able-bodied participants. Further testing is needed to determine if CUFF feedback can accelerate the mastery of myoelectric control or benefit specific patient sub-groups.

6.1 Introduction and Motivation

Limb loss is a condition that affects millions of people. Globally, 57.7 million people lived with traumatic limb loss in 2017 (McDonald et al. 2021), with the prevalence of amputation due to trauma/cancer in the United States alone expected to reach 1.35 million by 2050 (Ziegler-Graham et al. 2008). It is estimated that thirty percent of current people with amputation have lost an upper limb (https://u.osu.edu/wheelbarrows/upper-limb-injury-statistic 2024). That includes the hand, forearm, and upper arm. Upper limb loss can be especially life altering due to the dependence on bimanual operations for many activities of daily living, such as recreational activities and sports. Despite recent efforts to augment the lost motor function of the hand through active prostheses (Piazza et al. 2019), discontinued or reduced use of prostheses continues at high rates (Salminger et al. 2022; Resnik et al.

F. Barontini, *Wearable Haptic Devices for Realistic Scenario Applications*,
Springer Series on Touch and Haptic Systems,
https://doi.org/10.1007/978-3-031-70539-7_6

2019). The lack of proper restoration of touch-related sensory feedback is a severe drawback and may play a significant role in the reason for rejection of a prosthesis (Salminger et al. 2022). In addition, a residual limb with intact sensation often reveals itself to be more functional than an insensate prosthesis (Lundborg and Rosen 2001). Thus, sensation plays an important role in integration and embodiment of a prosthetic arm for the user. The current research to address loss of sensation in individuals with limb loss consists of the restoration of the haptic channel through invasive and non-invasive techniques. Invasive feedback can be provided with electrodes implanted surgically, relaying information directly to the nervous system (Navarro et al. 2005; Svensson et al. 2017). Non-invasive techniques rely on wearable haptic feedback devices. The latter solutions, referred to as supplementary feedback in this paper, convey touchmediated information through skin-stretch (Rossi et al. 2018), pressure (Tognetti et al. 2005; Barontini et al. 2021), vibration or electrical stimulation (Fani et al. 2019; Kaczmarek et al. 1991). Several types of information can be delivered through a supplementary feedback loop, including grip/grasp force, hand aperture, and contact (Sensinger and Dosen 2020). Survey results of upper limb prosthesis users suggest that a top priority for input in the feedback loop is the prosthetic hand's grip force magnitude (Markovic et al. 2018; De Nunzio et al. 2017; Ninu et al. 2014). Prosthesis users worry about over-squeezing and hurting others or breaking fragile objects (Graczyk et al. 2018). Previous studies have investigated modulation of grasp force using vibrotactile, electrotactile, mechano-tactile, and augmented reality feedback (Stephens-Fripp et al. 2018; Godfrey et al. 2018). In a thorough review of these studies by Stephens-Fripp et al. (2018) that identify current disadvantages to these feedback mechanisms and opportunities to bring effective supplementary feedback closer to fruition, the authors discuss the need for different testing approaches and feedback mechanisms for each type of prosthetic hand, such as the novel SoftHand Pro (SHP). The unique compliant grasp of the SHP differs significantly from the rigid prostheses that have been previously used in evaluating artificial feedback, and it is currently unknown whether conventional feedback approaches perform similarly in a prosthesis with a compliant grasp. The SHP is a robust and functional myoelectric prosthesis that minimizes cost and weight with an underactuated design of a single motor and soft adaptable robotics technologies (Godfrey et al. 2018). Functionally, this design is unique because of its synergistic movement pattern that adapts to the shape of the object it is grasping. The SHP offers a versatile platform into which a novel haptic device can be integrated. Therefore, an upper arm squeezing force feedback prototype was designed to communicate directly with the SHP platform. The Clenching Upper-Limb Force Feedback Device (CUFF) (Barontini et al. 2024) provides force feedback stimuli similar to the sensation of getting one's upper arm squeezed by a blood pressure cuff. Ajoudani et al. (2014), tested the CUFF with able-bodied participants in a manufacturing setting and found an improvement in tele-operation force control. In another study (Godfrey et al. 2016), the CUFF was used with the SHP to investigate grasp force while lifting an object in two different conditions (with and without the force feedback). The results did not show any significant differences between the two conditions; however, the study was only performed on a small cohort of able-bodied participants, and the CUFF's motor control algorithm was further optimized. Specifically, the original CUFF's motors were proportionally mapped to the closure of the SHP, while the current CUFF has a logarithmic relationship between the hand closure and the rotation of the CUFF's motors. This change was made because in a previous study participants desired more feedback sensitivity during larger grasp forces when the SHP would be at risk of over squeezing a fragile object. This study designed a new task that required force modulation and tested whether the use of the CUFF improved participants' ability to modulate the grasp force of the SHP. Our hypothesis was that the participant's grasping motion of the robotic hand, during the target force task, would be more precise with the CUFF than without and more similar to the sound limb.

6.2 Technologies and Instrumentation

In this section, we present the technological setup used in our experiments. This includes all the equipment, devices, and technologies employed to conduct the study.

6.2.1 The Softhand Pro

The SHP (Godfrey et al. 2017) is a myoelectric prosthetic version of the Pisa/IIT SoftHand. The SHP is designed and made by the authors team. Some components of the SHP are bought from qbrobotics s.r.l., a company producing robotic hands and an industrial version of the Pisa/IIT SoftHand. The Pisa/IIT SoftHand combined compliance and synergy inspiration (Catalano et al. 2012; Santello et al. 1998) into an artificial soft hand with 19 °C of freedom (DoFs), 4 on each finger, and 3 on the thumb. The fingers are capable of flexion/extension as well as ab/adduction. The hand is actuated by a single DC motor (15 Watt Maxon DCX 22S), and a gearhead of 86:1. With this setup, the hand can provide up to 76 N of force in power grasp and 20 N in pinch grasp. Further information about the SHP is published on the Natural Machine Motion Initiative website (http://www.naturalmachinemotioninitiative.net/) (Della Santina et al. 2017). The SHP can easily adapt to the environment as all of the fingers conform to the shape of grasped objects which augments its grasping capabilities. In addition, the SHP interfaces with commercially available surface electromyography (EMG) sensors (Ottobock, Germany) to allow control of the hand. EMG sensors were placed on the flexors and extensors of the forearm to activate closing and opening of the SHP, respectively. Proportional myoelectric control of the SHP used a FCFS (first come first serve) approach which acts on (by opening or closing) the first electrode signal to go above a minimum threshold and is controlled by only that signal until it drops below threshold (De Nunzio et al. 2017). For the cohort with limb loss, the SHP was connected to the participant's personal socket through a commercial wrist (Quick Disconnect Wrist, Ottobock, Germany), or a custom molded socket fabricated for an earlier SHP research study (Godfrey et al. 2017). For the able-bodied participants,

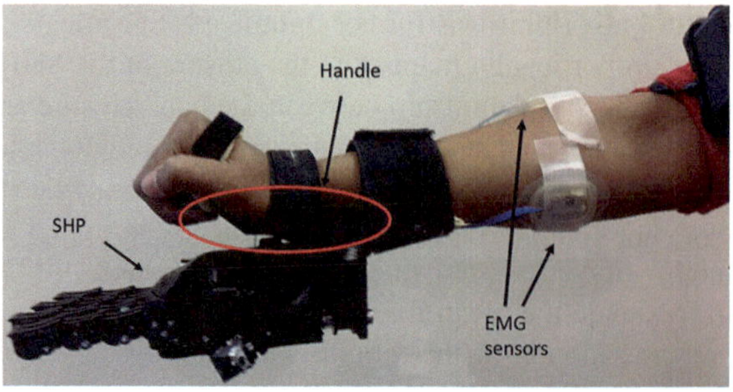

Fig. 6.1 The 3D printed Orthosis. The picture shows the SHP attached to a forearm adapter used by the able-bodied participants

the SHP was mounted on a one size fits all 3D printed orthosis (similar to a wrist cock up orthosis) that was secured on the user's forearm with a Velcro band (see Fig. 6.1).

6.2.2 *The Force Feedback Device CUFF*

As mentioned previously, the force feedback device used in this study is a reengineered version of the Clenching Upper-Limb Force Feedback (CUFF) (Schwarz et al. 2019). The redesign included a mechanical reduction of the device in terms of weight and dimension, and a different strategy for the motor control. Briefly, the CUFF device consists of a primary frame, which can be secured to the user's arm with two VELCRO straps, an actuation unit composed of two Maxon DC motors, powered by a 12 V battery, with a gearbox ratio of 64 : 1, and a belt used as an interface surface with the human body (see Fig. 6.2). The wearable device can provide distributed mechanotactile stimulation to the upper arm because the motors are programmed to rotate in opposite directions, tightening or loosening the band and increasing or decreasing the pressure on the arm. The user is instructed to associate the stimulus from the CUFF with the grip force of the SHP. To estimate the grip force exerted by the hand, the residual current (RC_{meas}) was calculated. The residual current is described as the difference between the estimated current and the real current absorbed by the motor of the SHP. For further detail see Mura et al. (2018). Mapping the information from the robotic hand into the feedback device requires an association between each value of the hand with a corresponding position value of the CUFF motor. A non-linear mapping was chosen to define the reference position (pos_{Cref}) of the motors. Two constraints were defined to identify the non-linear relation: when $RC_{meas} = 0$ also the $pos_{Cref} = 0$; and when $RC_{meas} = \frac{2}{3} RC_{max}$, $pos_{Cref} = \frac{1}{3} pos_{Cmax}$. So, the new reference position of the motors will be:

$$pos_{Cref} = \beta \log\left(1 - \alpha \frac{RC_{meas}}{RC_{max}}\right) pos_{Cmax} \tag{6.1}$$

Fig. 6.2 The Clenching Upper-Limb Force Feedback (CUFF). **a** On the left the CUFF device, and **b** a user wears the CUFF on their arm

The constraints were imposed by choosing the correct value for the constants: $\alpha = 0.9510$, and $\beta = -0.3317$. The values of the constant were extracted using the non-linear models provided by the Curve Fitting Toolbox in MATLAB. C_{max} position was set at 800 ticks from the zero position as a result of testing that determined this position to be the maximum position well tolerated by a group of individuals.

6.2.3 Instrumented Cylinder

As a means to assess the grip forces generated by the SHP, a cylindrical object (height 12 cm and diameter 7.5 cm) consisting of two hemicylindrical blades, was fabricated using 3D printing technology and each blade was instrumented with a 6-axis load cell (Nano 25, ATI Industrial Automation, Apex, NC, USA). Load cell data from this instrumented cylinder was collected with a 12 channel Analog/Digital board with USB communication (Model 8012, National Instruments, Austin, TX, USA) and sampled at 100 Hz using a simple control program (LabVIEW, National Instruments). Data were analyzed with a MATLAB script file (MATLAB 2019a, MathWorks Inc., Natick, MA, USA).

6.3 Methods and Experimental Protocol

In this section, we present the study schedule and the experimental protocol adopted during the clinical trial. This comprehensive overview includes the timeline of the study, detailed descriptions of each phase, and the specific procedures followed to ensure consistency and reliability in the data collection process.

6.3.1 Research Participants and Study Schedule

Two cohorts participated in this study. The first cohort was comprised of five participants with trans-radial limb loss (mean age 43.20 ± 16.79 years, 4 males/1 female, Table 6.1). The second cohort was comprised of 19 able-bodied participants (8 males, mean age 39.75 ± 8.87 years; 11 females, mean age 32 ± 11.79 years). The sample size was not determined from a power analysis. The number of participants in each cohort was the maximum number of participants that could be recruited and participate in data collection in the three-month period that was dedicated to this aspect of the project. This sample size also aligned with other prosthetic studies of these populations. Both groups were tested at Mayo Clinic in Rochester (MN, USA) using the SHP in conjunction with the CUFF. Eligibility criteria for the cohort with limb loss included individuals with a history of elbow disarticulation, trans-radial or trans-humeral limb loss ≥6 months prior to study enrollment, aged ≥18 years, and no prior experience utilizing the CUFF or any other force feedback device. Exclusion criteria included a clinical history of brachial plexopathy, cervical radiculopathy or polyneuropathy, orthopedic or joint impairments that would severely limit upper limb function, central nervous system disease limiting upper limb function, significant rigidity of the upper limb, medication use that might limit sensory or motor functions, and psychiatric or cognitive limitations that would prevent the ability to perform the study tasks.

For the cohort with limb loss, familiarization and data collection occurred over two consecutive days. The first day was dedicated to the familiarization with the SHP and the feedback CUFF as well as collection of two clinical assessments, the results of which are not reported in this work (Activities Measure for Upper Limb Amputee (AM-ULA), and the Jebsen Taylor Hand Function Test). The second day was devoted to biomechanical assessments of user interactions with the prosthetic and haptic feedback system and collection of grip force data using an instrumented cylinder. A sample of able-bodied participants aged ≥18 years, and without prior experience

Table 6.1 Demographics of the participants with limb loss. MP indicates Myoelectric Prosthesis, BP indicates Body-Powered

Participant	Age at time of testing (yrs)	Years Since Amputation (at time of study)	Side of Limb Loss	Previous Hand Dominance	Gender	Participant owned Prosthesis Type
1	31	7	L	R	M	BP* Hook
2	31	18	R	L	M	BP Hook
3	67	24	R	Unknown	F	Tridigit MP*
4	32	3	L	R	M	BP Hook
5	55	2	R	R	M	Multigrasp MP

with the force feedback CUFF the myoelectric SHP, was also recruited. Exclusion criteria were the same as for the group with limb loss, with the additional criteria that participants must be able to learn to effectively control myoelectric sensors within a one-hour timeframe. This additional constraint was included so that all interventions and assessments could be completed within a three-hour timeframe. A three-hour timeframe was chosen to allow for enough time to learn to control the SHP, complete the study tests, and allow for rests within the study session to avoid muscular fatigue. Participants were given the opportunity to familiarize themselves with the device in the first hour (approximately 35 min of myoelectric control training of the SHP without the CUFF, and approximately 25 min of training with both the SHP and CUFF). The remaining two hours were allocated for biomechanical assessments. If at the end of the first hour of familiarization the participant was unable to modulate grip force, testing with that participant was discontinued due to inadequate myoelectric control. Two male participants (aged 61 and 81) were not accrued after being recruited due to inadequate myoelectric control.

6.3.2 *Experimental Protocol*

The study protocol was approved by the Mayo Clinic Institutional Review Board (IRB), and all the participants gave written informed consent before enrollment in the study.

Familiarization

Once the SHP and socket or orthosis were donned by the participant, the gain settings for the myoelectric control were optimized. Participants of both cohorts familiarized themselves with control of the SHP (without the CUFF) for a period of 35 min. Familiarization included picking up and stacking objects including blocks and collapsible drinking cups, as well as picking up and placing down objects of different sizes, shapes, and materials. For the group with limb loss, additional familiarization (in addition to the 35 min) involved activities of daily living such as donning and doffing a shirt/sock, tying a shoe, writing, using scissors, picking up feeding utensils, turning paper, and picking up small objects such as paper clips and coins. Then a familiarization period with the addition of the force feedback CUFF occurred (approximately 25 min). Participants were instructed to focus on the sensation given by the CUFF during a range of grip forces from maximal grip to a very light grip. During familiarization and testing, participants were encouraged to take self-timed breaks and were also required to rest if they were showing symptoms of fatigue. All training and data collection phases were conducted by a licensed physical therapist (MVS).

6.3.2.1 Experimental Set-up

The participants were comfortably seated at an adjustable height table with the CUFF placed on the upper arm, and the cylinder in front of them. It should be noted that due to the design of the CUFF, it could be placed on any cylindrical body part such as the forearm or upper arm. To be consistent and not interfere with myoelectric sensor placement, this study placed the CUFF on the upper arm. The calibration of the CUFF involved an auto-adjustment of the fabric belt around the participant's arm in which the fabric was tightened until the two motors reached the stall condition and then release leaving the absorbed current close to zero. This process guarantees the complete contact of the belt on the arm. Therefore, the length of the belt is individualized to the anthropometric measurement of the participant's arm, but the starting pressure (close to zero) on the arm is standardized between participants. The cylinder was affixed to the table at a self-selected distance from the table edge (ranging from 10–15 cm from the table edge), on the side of the arm being tested Fig. 6.3. For the able-bodied cohort, the testing arm was the dominant arm. For the group with limb loss, the testing arm was the side of limb loss. The cylinder was rotated so that one half of the cylinder (one hemicylindrical blade) was in contact with the SHP's index and middle finger, and the other half was in contact with the opposing thumb during testing when gripping the cylinder from the top (see Fig. 6.4). The exact positioning of the cylinder was marked and repeated for both testing conditions (with and without CUFF feedback). A three-finger grasp was chosen to guarantee the best trial to trial repeatability. Due to the synergistic grasp and the shape of the instrumented object (cylindrical), other grasps caused the fourth and fifth fingers to slip off the cylinder during the grasp, rotating the cylinder, thus reducing trial to trial repeatability. Able-bodied participants supported an additional 520 g of mass distal to their native hand,

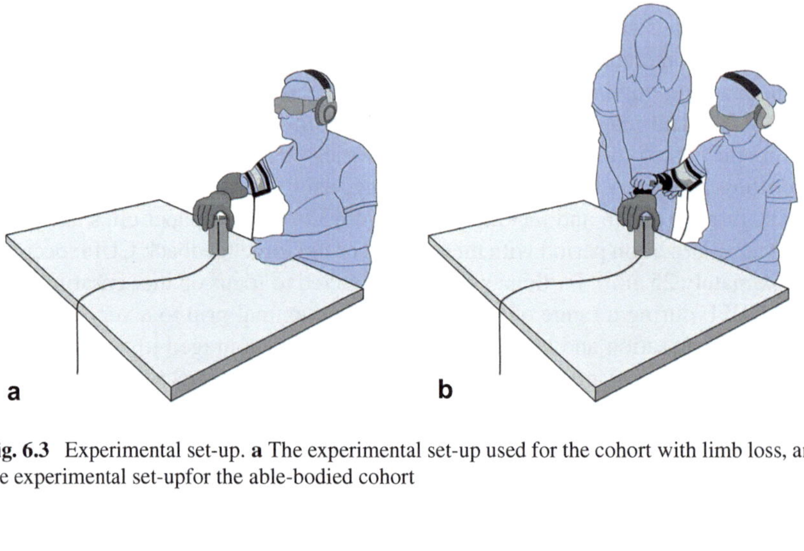

Fig. 6.3 Experimental set-up. **a** The experimental set-up used for the cohort with limb loss, and **b** the experimental set-upfor the able-bodied cohort

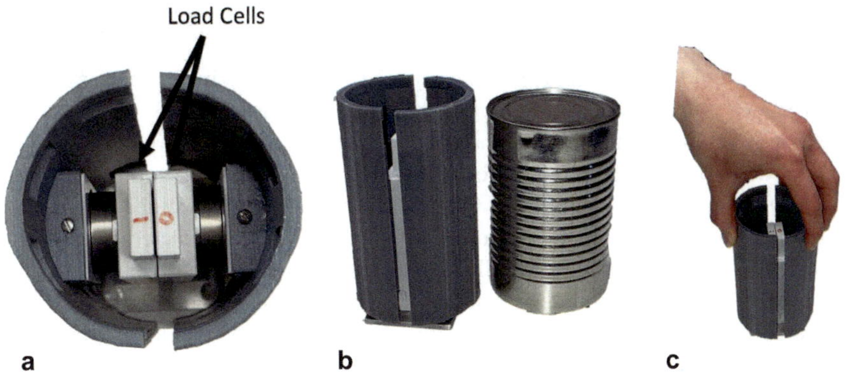

Fig. 6.4 Instrumented Cylinder. **a** From the left, view of the load cells from the top, **b** the cylinder next to a 400 g can, to understand the dimensions next to a functional object, and **c** an example of the grasp executed during the experiments, with the index and middle finger contacting one hemisphere of the cylinder and the thumb contacting the other hemisphere

which was the weight of the orthosis used to mount the SHP. In addition, the test required several minutes of consecutive static shoulder abduction, and the duration of the experimental protocol was 3 h. Therefore, manual support was provided to the able-bodied participants' forearms to prevent shoulder musculature fatigue and musculoskeletal overuse injury. A mobile arm support (i.e., gravity compensator) did not allow enough degrees of freedom to properly grip the cylinder with the SHP while still allowing for relaxation of the shoulder musculature; therefore, manual support was given by MVS. The physical therapist only provided support and reminders to allow her to support the weight of the arm; she did not provide any manual or verbal cues that would influence performance during testing. Manual forearm support was not provided to the participants with limb loss since they had to lift only the SHP which was similar in weight to their owned prostheses and the mass of the SHP did not extend beyond that of a native hand.

Training for the targeted force

Prior to data collection, participants were trained to grip with a targeted moderate amount of force (less than the maximal force generated by the SHP and more than the minimum grip required to activate the CUFF). This training provided the participants with an understanding of the stimuli provided by the force feedback CUFF related to the grip of the SHP.

1. *Quantifying the minimum grip force to activate the CUFF*. The participants were instructed to slowly squeeze the cylinder and stop squeezing the moment they felt the first stimulus exerted by the CUFF on the arm. The minimum force grasp was repeated three times and the applied force of the two loads cells was recorded. The

median value of peak force applied in all the trials was defined as the minimum grip force.

2. *Quantifying the maximum grip force of the SHP*. The participants were instructed to squeeze the cylinder as hard as they could and hold the grip for 3 s. The maximal grasp was repeated three times and the peak of the force of the two load cells was recorded. The median of multiple trials was defined as the maximum grip force.
3. *Training to modulate grip force*. Once the participants were trained to sense the minimal and maximal stimulus of the CUFF, they were instructed to gradually squeeze the cylinder with the SHP to receive a moderate amount of stimulus on their arm, thus producing a moderate grasp force. Performing a moderate amount of grasp force (range) rather than a specific grasp magnitude (point) was the verbal cue since the SHP's synergistic grasp creates a residual current that varies between trials (of the same participant) and this current tells the CUFF how much to squeeze. Based on the displacement of the CUFF fabric (caused by rotation of the motors), the examiners provided verbal feedback to indicate if the training trial was "too much" (when they reached the maximum force), "not enough" (when the grasp was not forceful enough to activate the CUFF), or "good" (when the grip was larger than minimum but less than maximum grip force). Following 10 trials for which the participant received the verbal feedback of "good", the training was complete. The forces, registered by the load cell, were recorded during the training. Video was also recorded during training. Following data collection on a subsequent day, video was reviewed and the trials that were "good" were identified so that the median moderate grip force of the "good" trials could be calculated.

6.4 Data Collection

Headphones and non-prescriptive glasses were donned by the participants for data collection. The headphones played pink noise (https://mynoise.net/NoiseMachines/whiteNoiseGenerator.php) to eliminate auditory feedback about the grip produced by the actuators of the SHP and the CUFF. Pink noise delivers less intensity in the higher frequencies and more intensity in the lower frequencies, making it more comfortable for the human ear than white noise and still effective at blocking sound. The lenses of the glasses were modified to reduce vision so that participants could not determine the positioning of the fingers of the SHP around objects. A layer of thin, opaque foam was adhered to the lenses to obstruct participants' view. Prior to donning the headphones and glasses, participants were instructed to squeeze the cylinder with moderate force as they had done during training. They were told that two sets of five repetitions per condition would be completed with 2 min of rest between sets. During rest, the headphones and glasses were removed. Data from the instrumented cylinder was collected at 100 Hz. During data collection, examiners noted, not surprisingly, that there were differences, among the able-bodied participants, in their ability to use a myoelectric prosthesis. Video of the data collection was reviewed on

Table 6.2 The possible testing orders accomplished by the able-bodied participants during the experiment

Testing order	First testing condition	Second testing condition	Third testing condition
1	SHP + CUFF	Sound or Controlateral Limb	SHP
2	SHP	Sound or Controlateral Limb	SHP + CUFF

a subsequent day by MVS and the participants were subjectively categorized on a 1–3 scale of myoelectric control/skill. Participants were assigned a 3 if they consistently demonstrated the ability to contract wrist flexor muscles for SHP grasping and wrist extensors for SHP opening without co-contraction, a 2 if they began the data collection requiring moderate cuing to relax certain muscle groups in order to avoid co-contraction but improved over the course of the training, and a 1 if they required maximal cuing throughout the training and co-contraction was observed frequently even during testing.

Testing order

For the able-bodied participants, block randomization (in groups of six) of the testing order of condition (utilizing the CUFF versus not utilizing the CUFF) was performed to assure that equal number of participants performed each testing order. Table 6.2 shows the two possible testing orders. Ten participants used the CUFF first (6 male/4 female) and 9 participants used the CUFF last (2 male/7 female).

The sound limb or the controlateral limb testing was done in between conditions with the SHP to provide musculoskeletal rest for the limb performing myoelectric control of the SHP. For the participants with limb loss, the first participant began the test while not wearing the CUFF. The subsequent testing alternated between wearing the CUFF first and not wearing the CUFF first, etc. The exact positioning of the cylinder was marked between conditions so that the rotation of the cylinder could be standardized.

6.5 Data Analysis

The data recorded were analyzed with functional Principal Component Analysis (fPCA). Briefly, the fPCA is a statistical method that allows comparison between motions that occur in nature, such as human movement (Averta et al. 2019). In human movement, variability between repetitions of the same task naturally exists. This variability is larger in humans with impairments such as hemiparesis due to stroke (Schwarz et al. 2019) and missing sensation due to limb loss. The goal of rehabilitative interventions is to reduce this variability which can be quantified with fPCA over the entire motion of the task. In this study, we used the fPCA to compare differences in precision, during a grasping motion, with or without the use of CUFF feedback. To avoid the inclusion in this analysis of undesired features (an accidental grasp before the trial was started, or pressure on the load cells due to resting the hand with a small amount of pressure on the top of the cylinder before grasping), due to misalignments in time, the following pre-processing procedure was performed to all the data recorded from instrumented cylinder task:

1. The signal was manually segmented to remove the portion of the recording in which the participant was not actively performing the task. 50 frames before and after each repetition were included.
2. A resample of the signal was executed to have a standardized length of 400 samples for each trial.
3. A time warping technique synchronized (in time) all the signals of the data. This procedure, known in the literature as Dynamic Time Warping (DTW), identifies the optimal time-shift and time-stretch of one signal with regard to another. DTW, described in detail by Averta et al. (2019), was necessary to mediate the ten repetitions and to avoid misalignment in this data.

In this work, we quantified differences in precision of force during an entire grasping motion, with or without the use of CUFF feedback. The main hypothesis was that with the stimuli provided by the CUFF, the subjects would be able to grasp in a more natural way (similar to the sound limb testing condition), compared to the no feedback condition. Aligned with this hypothesis, given a specific number of fPCs used to reconstruct the signal, the explained variance would be lowest in condition of no CUFF feedback. If the first functional component achieves 100% of the explained variance (of the grasping force motion), then the grasp would be considered perfectly precise. However, human movement is never perfectly precise and involves corrections during a motion as well as variance between repetitions. fPCA allows us to speculate that if the explained variance is high there were few or no corrections in the grasp action, and that the participant was able to grasp with similar force patterns from trial to trial. If there is the necessity to add more components to explain the signal, then there are likely more corrective actions to obtain the targeted force and more variability between trials.

6.6 Results

As expected, and for both cohorts, the sound limb condition resulted in higher first functional principal components (95%—participants with limb loss; 89%—able-bodied cohort) than any condition with the SHP. For the participants with limb loss, use of the CUFF resulted in the first functional principal component explaining 83%

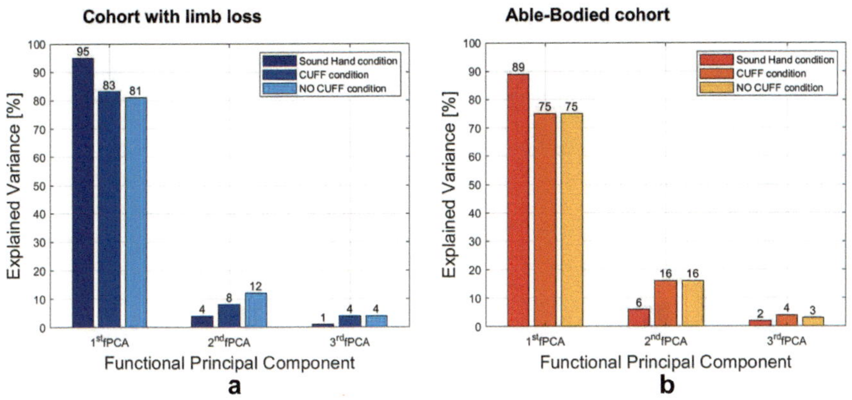

Fig. 6.5 Explain variance of the first three fPCA for the entire cohort of the participant with limb loss (on the left), and able-bodied participants (on the right)

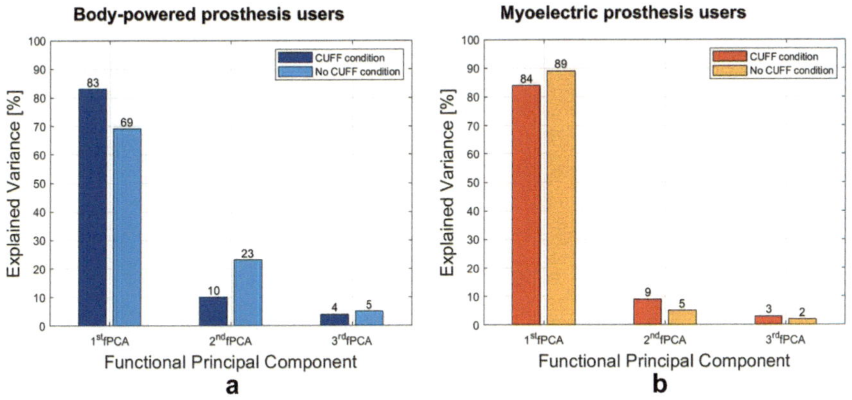

Fig. 6.6 Explained variance of the first three principal components for the sub-group of body-powered prosthesis users (on the left), and the sub-group of the myoelectric prosthesis users (on the right)

of the variability compared to 81% without the CUFF (Fig. 6.5). Because participants used different types of prostheses in their daily life (Table 6.1), the limb loss participants ($N = 5$) were categorized into two subgroups: body-powered ($N = 2$) and myoelectric prosthesis users ($N = 3$). Analysis by subgroup indicated that the participants who regularly used a body-powered prosthesis, when using the CUFF, reached 83% of explained variance compared to 69% of explained variance when the CUFF was not used. However, experienced users of myoelectric prostheses had percentages of explained variance of 84.5% and 89% with and without the CUFF, respectively see Fig. 6.6. For the able-bodied cohort, there was no substantial difference in percentage of explained variance between the two conditions with and without the CUFF (Fig. 6.7). However, the sub-group of five participants, who were

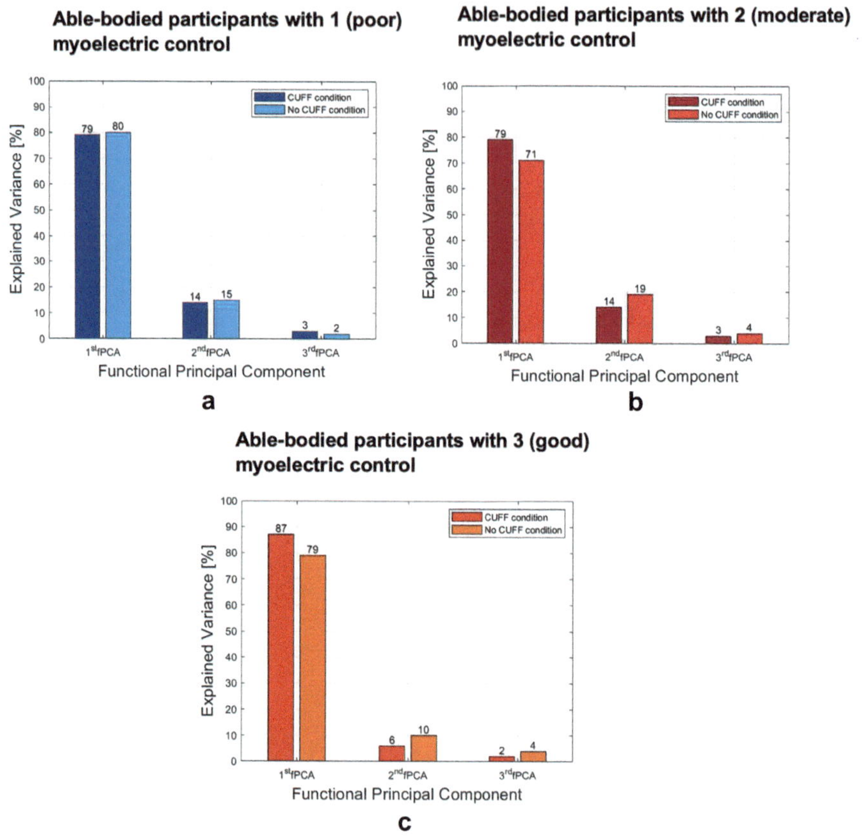

Fig. 6.7 Explained variance for the sub-group of the able-bodied participants by subjective rating of myoelectric control skill. On the top left are the results of the sub-group with "poor" myoelectric control who required maximal cues to minimize co-contraction of their wrist muscles. On the top right are the results of the sub-group that needed "moderate" cues but improved during training; and on the bottom the sub-group that demonstrated "good" myoelectric control without co-contraction of the wrist muscles

subjectively rated to have poor myoelectric control skills (score of 1), obtained a percentage of explained variance of 79% and 80% for the condition with and without the CUFF, respectively. For the moderate myoelectric control skills (score of 2) subgroup, the explained variance was 79% with the CUFF and 71% without the CUFF. For the good myoelectric control skills (score of 3) subgroup, the explained variance was 87% with the CUFF compared to 79% without the CUFF (see Fig. 6.7).

To assist with the interpretability of the fPCA, Fig. 6.8 is provided which depicts individual force curve data for three participants; a participant with limb loss who typically uses a body powered prosthesis and two able-bodied participants who began

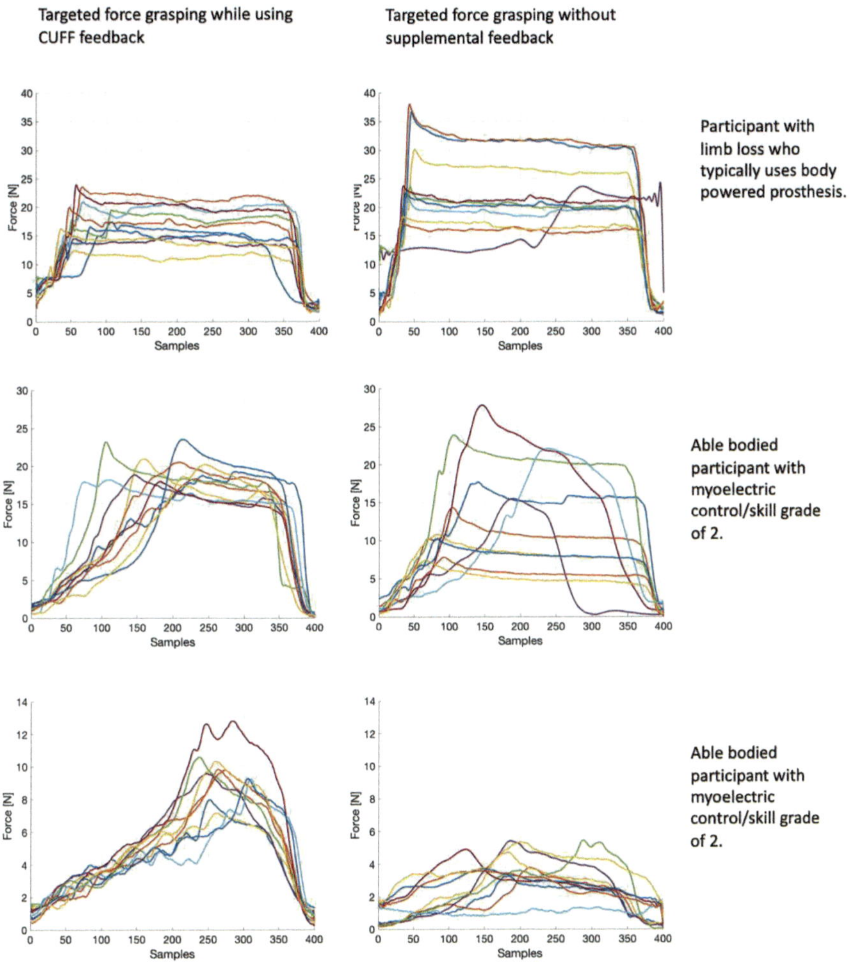

Fig. 6.8 Individual forces curves for one participant with limb loss and two able-bodied participants. Participants completed 10 trials of a targeted force grasp with CUFF feedback (left column) and without supplemental feedback (right column). The top row is the force curves from a participant with limb loss who typically uses a body powered prosthesis. The bottom two rows are force curves from two able-bodied participants who were subjectively graded as a 2 for the myoelectric control/skill category. Each trial on the graphs represents a full grasping cycle

the study requiring moderate cuing to relax certain muscle groups in order to avoid co-contraction but improved over the course of the training (myoelectric control/skill grade of 2). In these three examples, the trials with the feedback from the CUFF had visually less variability between trials. This can be seen in the magnitude of the force as well as the shape of the force curves.

6.7 Discussion

This study tested the ability of the CUFF, which provided SHP grip force feedback, to improve the participants' grasp force control. To add context to these findings, the grasp of the sound limb was also tested, and as expected, was shown to have more grasping precision than the SHP. The precision of the grasping motion (quantified with fPCA) did not differ between the CUFF and no CUFF conditions for either those with limb loss or the able-bodied cohort. Although grasping precision differences were not seen when the cohorts were analyzed as a whole, the subgroup of participants with limb loss who regularly use body powered prostheses and the subgroups of able-bodied participants who demonstrated moderate and poor skill of myoelectric control demonstrated improved precision of the grasping motion (as measured by fPCA) when the CUFF was used. Participants of both cohorts had better grasp precision with their sound limb (as measured by fPCA), as expected. In Averta et al. the input kinematic data were analyzed with the fPCA to investigate upper limb movement following stroke (Schwarz et al. 2019), other examples of using fPCA to investigate upper limb motions have not been found in the literature. This study's approach of inputting force data during a grasping task into fPCA adds to the limited body of work utilizing fPCA to compare upper limb movement patterns. FPCA appears to be a promising tool to differentiate groups of individuals who have varied prosthetic experience, coordination, or myoelectric control. According with the results, for participants with limb loss who had the least amount of experience using myoelectric control (body powered prosthesis users), the force feedback seemed to improve grasping precision. A plausible explanation for this is that humans develop internal models through experiences and these internal models allow them to function efficiently using feed forward approaches (Sensinger and Dosen 2020). In a recent review of the literature, Sensinger and Dosen (2020) present a comprehensive overview of motor learning principles related to both natural and supplemental closed feedback loops; they propose how this knowledge should guide data interpretation and future research and development related to upper limb prostheses. Sensinger and Dosen state that *"when subjects have not yet received extensive training, (supplemental) feedback is useful—both to develop internal models as well as execute real-time corrections. Over time, however, as participants develop better internal models, the usefulness of (supplemental) feedback for real-time corrections may fade"*. Training allows humans to perceive and interpret incidental sources of information. The prosthetic user receives vibratory and auditory cues from the motor. Prosthetic users become highly reliant on the high-fidelity visual information of the prosthetic hand's movement. Training also allows users to understand the graded output of their forearm muscle contractions. While our study design reduced the visual and auditory information about the prosthetic hand, participants were still able to benefit from vibratory cues caused by the motor and internal models about muscle contractions. Likely strong internal models and use of incidental feedback allowed the myoelectric prosthetic users to perform well even without the use of the CUFF. There was also no benefit to using the CUFF for the sub-set of able-bodied participants who were subjectively graded as a 1 (indicating that throughout the course of the experiment they had poor myoelectric control). Our able-bodied participants had a very short time to familiarize themselves with myoelectric control and this sub-set of individuals may have needed more time to develop coordination of the forearm muscles before the effect of the force feedback could be tested. In summary, the value of supplemental feedback is reduced when already-available incidental feedback exists along with a well-developed internal model. While it is important not to translate findings from the able-bodied cohort into clinical applications for those with limb loss, able-bodied cohorts with simulated functional limitations are sometimes studied when sampling with proper power would be too time and resource intensive (Musolf et al. 2021; Fu et al. 2019). Findings from an able-bodied cohort can then help refine research questions for future clinical trials for persons with limb loss, such as the work done by Fu et al. who demonstrated that inter-limb training improved participants' ability to grasp a targeted force (Fu et al. 2019). It is possible that the able-bodied cohort in this study was representative of patients with a relatively new limb loss condition since both are naïve to using a prosthesis. If grasp force feedback for the SHP does improve precision in persons with new limb loss, then technology such as the CUFF would have value in rehabilitation settings. Findings that the CUFF benefited persons with limb loss who have less myoelectric control experience support the results from the able-bodied cohort. The CUFF may assist myoelectric control users of the SHP become proficient more quickly. However, the sample size of participants with limb loss was small, therefore interpretations from these data should be made cautiously, and future studies should have larger sample sizes of persons living with limb loss. Likewise, future research should investigate whether current grasp force feedback technologies are useful only during early learning stages or whether there is a long-term benefit and therefore should be integrated into sockets. One of the limitations of this study design was that the task of grasping the top of the cylinder did not also include lifting the cylinder. A large majority of functional tasks involve a grasp and a lift, and therefore the study task was not functional. Additionally, in the real-world setting, individuals would not have had vision and hearing blocked, therefore, our approach was so constrained that it does not translate to the natural free-living environment. These methods were chosen because the unique design of the SHP (single tendon, flexion synergy, single motor) requires an incremental investigation of supplementary feedback that starts with demonstration that the provided feedback interface is effective in transmitting the desired grasp force information (Sensinger and Dosen 2020). The next step in designing and testing supplementary feedback for the SHP will require addressing the limitations of the current study; (1) measuring whether supplementary feedback benefits the user even with incidental feedback of vision and hearing, (2) testing tasks that are more representative of functional activities, and (3) allowing participants to use feedback over a longer time period and in the home setting as is currently being done by others (Graczyk et al. 2018). Further, since the CUFF feedback demonstrated improved precision during grasping in sub-groups of our two cohorts, understanding if supplemental feedback is beneficial to accelerate mastery of myoelectric control or with certain patient sub-groups is also needed.

6.8 Conclusion

This study demonstrated that the supplementary force feedback provided by the CUFF was beneficial to modulating the grasp force of the SHP with precision, however the in-lab testing protocol may not translate to the real-world and the CUFF may not benefit those who already have extensive experience with myoelectric control. Future development of supplementary force feedback for the SHP should include functional tasks in free-living environment, no restrictions to incidental feedback that is available in the real world, and longer-term use of the supplemental feedback.

Acknowledgements We would like to thank all of our research participants who dedicated their time to improving upper limb protheses through research, and our essential study coordinators, Tyson L. Scrabeck and Julie Block.

References

Ajoudani A, Godfrey SB, Bianchi M, Catalano MG, Grioli G, Tsagarakis N, Bicchi A (2014) Exploring teleimpedance and tactile feedback for intuitive control of the pisa/iit softhand. IEEE Trans Haptics 7(2):203–215

Averta G, Valenza G, Catrambone V, Barontini F, Scilingo EP, Bicchi A, Bianchi M (2019) On the time-invariance properties of upper limb synergies. IEEE Trans Neural Syst Rehabil Eng 27(7):1397–1406

Barontini F, Catalano MG, Fani S, Grioli G, Bianchi M, Bicchi A (2024) The cuff, clenching upper-limb force feedback wearable device: Design, characterization and validation. IEEE Trans Haptics

Barontini F, Catalano MG, Grioli G, Bianchi M, Bicchi A (2021) Wearable integrated soft haptics in a prosthetic socket. IEEE Robot Autom Lett 6(2):1785–1792

Catalano MG, Grioli G, Serio A, Farnioli E, Piazza C, Bicchi A (2012) Adaptive synergies for a humanoid robot hand. In: 12th IEEE-RAS international conference on humanoid robots (Humanoids 2012). IEEE, pp 7–14

De Nunzio AM, Dosen S, Lemling S, Markovic M, Schweisfurth MA, Ge N, Graimann B, Falla D, Farina D (2017) Tactile feedback is an effective instrument for the training of grasping with a prosthesis at low-and medium-force levels. Exp Brain Res 235:2547–2559

Della Santina C, Piazza C, Gasparri GM, Bonilla M, Catalano MG, Grioli G, Garabini M, Bicchi A (2017) The quest for natural machine motion: an open platform to fast-prototyping articulated soft robots. IEEE Robot Autom Mag 24(1):48–56

Fani S, Di Blasio K, Bianchi M, Catalano MG, Grioli G, Bicchi A (2019) Relaying the high-frequency contents of tactile feedback to robotic prosthesis users: design, filtering, implementation, and validation. IEEE Robot Autom Lett 4(2):926–933

Fu Q, Shao F, Santello M (2019) Inter-limb transfer of grasp force perception with closed-loop hand prosthesis. IEEE Trans Neural Syst Rehabil Eng 27(5):927–936

Godfrey SB, Bianchi M, Bicchi A, Santello M (2016) Influence of force feedback on grasp force modulation in prosthetic applications: a preliminary study. In: 2016 38th annual international conference of the ieee engineering in medicine and biology society (EMBC). IEEE, pp 5439–5442

Godfrey SB, Bianchi M, Zhao K, Catalano M, Breighner R, Theuer A, Andrews K, Grioli G, Santello M, Bicchi A (2017) The softhand pro: Translation from robotic hand to prosthetic prototype. In: Converging clinical and engineering research on neurorehabilitation II: proceedings of the 3rd international conference on neurorehabilitation (ICNR2016). Springer International Publishing, Segovia, Spain, pp 469–473

Godfrey SB, Zhao KD, Theuer A, Catalano MG, Bianchi M, Breighner R, Bhaskaran D, lennon R, Grioli G, Santello M, Bicchi A, Andrews K (2018) The SoftHand Pro: Functional evaluation of a novel, flexible, and robust myoelectric prosthesis. PloS One 13(10):e0205653

Graczyk EL, Resnik L, Schiefer MA, Schmitt MS, Tyler DJ (2018) Home use of a neural-connected sensory prosthesis provides the functional and psychosocial experience of having a hand again. Sci Rep 8(1):9866

https://u.osu.edu/wheelbarrows/upper-limb-injury-statistic

Kaczmarek KA, Webster JG, Bach-y-Rita P, Tompkins WJ (1991) Electrotactile and vibrotactile displays for sensory substitution systems. IEEE Trans Biomed Eng 38(1):1–16

Lundborg G, Rosen B (2001) Sensory substitution in prosthetics. Hand Clin 17(3):481–488

Markovic M, Schweisfurth MA, Engels LF, Farina D, Dosen S (2018) Myocontrol is closed-loop control: incidental feedback is sufficient for scaling the prosthesis force in routine grasping. J Neuroeng Rehabil 15:1–11

McDonald CL, Westcott-McCoy S, Weaver MR, Haagsma J, Kartin D (2021) Global prevalence of traumatic non-fatal limb amputation. Prosthet Orthot Int 45(2):105–114

Mura D, Barbarossa M, Dinuzzi G, Grioli G, Caiti A, Catalano MG (2018) A soft modular end effector for underwater manipulation: a gentle, adaptable grasp for the ocean depths. IEEE Robot Autom Mag 25(4):45–56

Musolf BM, Earley EJ, Munoz-Novoa M, Ortiz-Catalan M (2021) Analysis and design of a bypass socket for transradial amputations. In: 2021 43rd annual international conference of the ieee engineering in medicine and biology society (EMBC). IEEE, pp 4611–4614

Navarro X, Krueger TB, Lago N, Micera S, Stieglitz T, Dario P (2005) A critical review of interfaces with the peripheral nervous system for the control of neuroprostheses and hybrid bionic systems. J Peripheral Nervous Syst 10(3):229–258

Ninu A, Dosen S, Muceli S, Rattay F, Dietl H, Farina D (2014) Closed-loop control of grasping with a myoelectric hand prosthesis: which are the relevant feedback variables for force control? IEEE Trans Neural Syst Rehabil Eng 22(5):1041–1052

Piazza C, Grioli G, Catalano MG, Bicchi AJAROC (2019) A century of robotic hands. Ann Rev Control Robot Auton Syst 2:1–32

Resnik L, Ekerholm S, Borgia M, Clark MA (2019) A national study of Veterans with major upper limb amputation: survey methods, participants, and summary findings. PloS One 14(3):e0213578

Rossi M, Bianchi M, Battaglia E, Catalano MG, Bicchi A (2018) HapPro: a wearable haptic device for proprioceptive feedback. IEEE Trans Biomed Eng 66(1):138–149

Salminger S, Stino H, Pichler LH, Gstoettner C, Sturma A, Mayer JA, Szivak M, Aszmann OC (2022) Current rates of prosthetic usage in upper-limb amputees-have innovations had an impact on device acceptance? Disabil Rehabil 44(14):3708–3713

Santello M, Flanders M, Soechting JF (1998) Postural hand synergies for tool use. J Neurosci 18(23):10105–10115

Schwarz A, Averta G, Veerbeek JM, Luft AR, Held JP, Valenza G, Bicchi A, Bianchi M (2019) A functional analysis-based approach to quantify upper limb impairment level in chronic stroke patients: a pilot study. In: 2019 41st annual international conference of the IEEE engineering in medicine and biology society (EMBC). IEEE, pp 4198–4204

Sensinger JW, Dosen S (2020) A review of sensory feedback in upper-limb prostheses from the perspective of human motor control. Front Neurosci 14:519626

Stephens-Fripp B, Alici G, Mutlu R (2018) A review of non-invasive sensory feedback methods for transradial prosthetic hands. IEEE Access 6:6878–6899

Svensson P, Wijk U, Björkman A, Antfolk C (2017) A review of invasive and non-invasive sensory feedback in upper limb prostheses. Expert Rev Med Dev 14(6):439–447

Tognetti A, Lorussi F, Bartalesi R, Quaglini S, Tesconi M, Zupone G, De Rossi D (2005) Wearable kinesthetic system for capturing and classifying upper limb gesture in post-stroke rehabilitation. J NeuroEng Rehabil 2:1–16

Ziegler-Graham K, MacKenzie EJ, Ephraim PL, Travison TG, Brookmeyer R (2008) Estimating the prevalence of limb loss in the United States: 2005 to 2050. Arch Phys Med Rehabil 89(3):422–429

Chapter 7
Toward Somatotopic Matching: The Wearable Integrated Soft Haptic Device (WISH)

Abstract This work presents a soft pneumatic feedback system designed with integrability and wearability as primary concerns. Modern active prostheses can help recover some motor functions lost with hand amputation, but they typically fail to restore sensory abilities, forcing individuals to rely predominantly on visual feedback. Few solutions have been effectively integrated into prostheses for real-world use. At the system's core, two soft pneumatic actuators are placed in contact with the subject's skin and inflated to provide pressure stimuli, representing the force exerted during hand grasping. We report on the design and characterization of the system, including behavioural experiments with able-bodied participants and one prosthesis user. Results from psychophysical, dexterity, and usability tests show that the system has the potential to restore sensory feedback in hand amputees and can be a useful tool for enabling correct modulation of force during grasping and manipulation tasks.

7.1 Introduction

The interaction of humans with the surrounding world is an extremely complex phenomenon, mostly enabled by our hands. These play an active role in the application of forces to external objects, while receiving a large amount of sensing information about the external world through the tactile channels such as objects' softness, dimensions, texture, temperature, and so on Katz et al. (2013). Unfortunately, this incredible biomechanical system can suffer irreversible injuries that prevent an effective implementation of the natural sensory-motor capabilities. In the most severe cases, these can lead to upper extremity amputation, a phenomenon that affects over 2 million people in US only,[1] with a significant increase of the average related healthcare costs (over 150K dollars) with respect to non-amputee population (Biddiss et al. 2007; Gurari et al. 2012).

The development of novel myo-electric (electromyography—driven) mechatronic prostheses (Piazza et al. 2019) obtained important results for the restoration of

[1] PhaseSpace, Inc., http://www.biotechpossibilities.com.

F. Barontini, *Wearable Haptic Devices for Realistic Scenario Applications*, Springer Series on Touch and Haptic Systems,
https://doi.org/10.1007/978-3-031-70539-7_7

the motoric components, lost with the amputation. However, there are no matching advancements in the sensory domain yet, to restore the role of the hands as information collectors through the haptic channel. This, in turn, translates also in a sub-optimal control performance in everyday use of the prosthesis (Culbertson et al. 2018), since the user has to rely only on the exploitation of visual feedback, which cannot always provide a satisfactory amount of information (Fraser et al. 1981).

Although, the restoration of the haptic channels can be implemented by relying on invasive systems that interface to the user nervous system (Oddo et al. 2016), such solutions are still very experimental. Therefore, much research focuses on alternative non-invasive techniques to restore the haptic channel on the amputates limb. The latter exploit sensory substitution through the usage of wearable haptic devices (Svensson et al. 2017). These convey touch-mediated information through e.g. skin-stretch, pressure, vibrations or electrical stimulation (see e.g. Rossi et al. 2018; Stanley and Kuchenbecker 2012; Clark et al. 2018). An ideal sensory substitution loop should provide (Kim and Colgate 2012) modality matching (MM), by delivering artificial cues that share the same sensory modality as the natural sensation, and somatotopic matching (SM), by delivering cues in the same place where they would naturally be felt. Although perfect SM is impossible to achieve on an amputee for obvious reasons, it was observed in Antfolk et al. (2012) that the sensation elicited by contacts at the stump level can be interpreted by the central nervous system as similar to the sensation perceived at the finger level. Ideally a device integrated inside a prosthesis could be able to apply haptic stimuli directly at the stump level, and may effectively overcoming the limitation of previous solutions and satisfy the somatotopic matching paradigm, without invasive solution. Studies focusing on discrete feedback, such as Clemente et al. (2015), placed pressure sensor thimbles on a prosthetic hand, and used a cuff to convey vibrational stimuli on the upper arm. In Fani et al. (2019) the authors used vibrational stimuli to convey physical information of the surface and first-contact cue. Antfolk et al. (2013) used five controlled pressure devices to transfer tactile input to various locations of the forearm. In an attempt to go a step further towards the realization of the ideal feedback, we present a novel wearable soft pneumatic force feedback, composed by two silicone chambers, used to exert pressures in specific stump sites. This solution, named Wearable Integrated Soft Haptic device (WISH), strives to achieve both good modality matching (MM) and somatotopic matching (SM). Indeed, according to Kim and Colgate (2012) the exertion of a pressure cue has the same modality of stimuli related to those naturally associated to grasping forces, while the application of these stimuli at the stump level provides a good somatotopic approximation, as demonstrated in Antfolk et al. (2012). Other studies tried to exploit soft robotic technologies to propose a MM and SM stimuli. In Huaroto et al. (2018), the authors present a soft pneumatic actuators able to provide tactile and vibration feedback. In Wijk et al. (2020), the authors present a passive device based on a glove with silicon bulbs in the fingertips. When they are pressed, the air is transferred via silicone pipes that reached actuators inside the prosthetic socket and gave pressure (mechanotactile feedback) on the skin corresponding with the phantom hand map zones in the residual limb. Despite the system obtained a positive rating, the stimuli provided were not meaningful. Possible reasons could be due to the fact

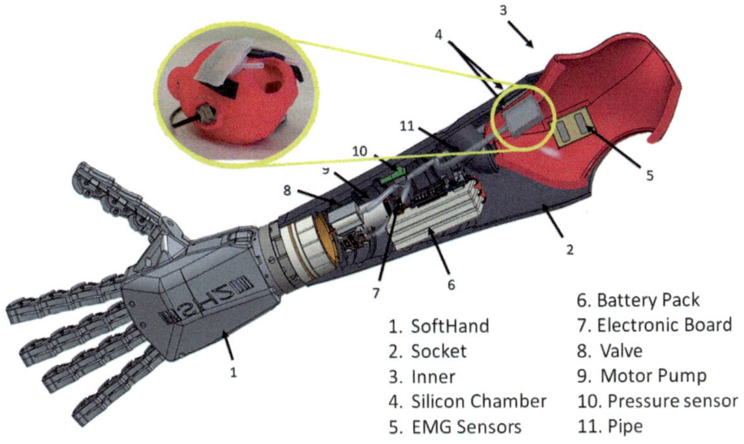

Fig. 7.1 Overview of the main components of the Wearable Integrated Soft Haptics device

that this is a passive feedback and it is not possible to amplify the signal. Moreover each stimulus was mapped in a single pressure stimulus, and the subject typically missed the perception of the overall pressure. The novelty of the WISH w.r.t. current solutions is the ability to amplify, modulate and convey a pressure stimulus that represents the overall force applied by the robotic hand on an object during the grasp phase. Furthermore it is worth mentioning that WISH can be easily integrated inside the inner socket (see Fig. 7.1).

7.2 The System Design

The Soft Haptic pneumatic system is composed by a mechanical actuation unit powered by a diaphragm pump, and a feedback interface composed by two silicone chambers. We choose a rectangular shape to design the chambers, which have to be integrated into the inner socket. In this manner, we aim at maximizing the surface available in the inner.

7.2.1 Material and Fabrication Method

The chambers are fabricated with a bio-compatible silicone rubber, in order to guarantee a safe contact with the skin. More specifically, we opted for the silicone Ecoflex 30,[2] a platinum-catalyzed silicone, because of its softness (10 psi, 100% Modulus),

[2] https://www.smooth-on.com/products/ecoflex-00-30.

Fig. 7.2 Fabrication process of the silicone chambers Fig. 7.2a. The numbers indicate the components of the mold used to create the chambers. On the right the whole chamber

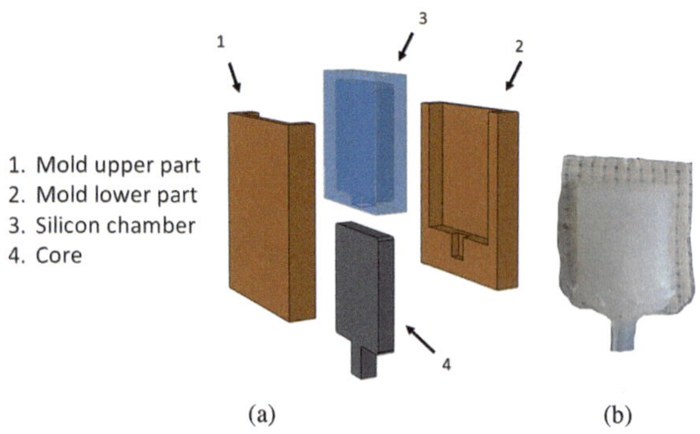

and robustness (Tensile Strength 200 psi, and Elongation at Break 900%) . The chambers were fabricated using a custom 3-part mold, shown in Fig. 7.2a. Finally it is made to incorporate a piece of non-elastic fabric (0.4 mm thick cotton 11-count aida cloth—10 squares per inch) and a second layer of Ecoflex was poured on it. The non-elastic fabric is used as constraint, in order to prevent the inflation of the chamber in both sides. Indeed, our goal is to maximize the inflation of the chamber in contact with the forearm skin. We 3D printed a mold to fabricate the air chambers, composed of an upper and a lower part, and a central core (see Fig. 7.2a). The fabrication procedure include the following steps: (1) we mixed the silicone Part A and Part B in equal portion (1A:1B by weight or volume) thoroughly for 3 min; (2) we put the rubber in a vacuum chamber (2–3 min at 29 inches of mercury) until the surface became clear and without bubble; (3) we poured the mix inside the mold, and we left it to cure for at least 4 h at ambient temperature. We took off the core (5); (6) we glued a silicone connecting pipe using a commercial adhesive silicone Sil-Poxy,[3] and we left it to cure at ambient temperature for 30 min; (7) the silicone chamber was positioned on a piece of non-elastic fabric (0.4 mm thick cotton 11-count aida cloth—10 squares per inch) and a second layer of Ecoflex was poured on it. Figure 7.2b show the final result.

7.2.2 The Pneumatic System

Figure 7.3 shows an overview of the complete haptic system, which is composed by two main subsystems: the feedback interface and the mechanical actuation unit.

The feedback interface is composed by two silicone chambers (5, 6 in Fig. 7.3) connected to the mechanical actuation unit with a silicone connecting pipe. The actuation unit is powered by a D220S diaphragm pump (by TSC Micropumps, (1) Fig. 7.3), with a size of 42 × 27 × 17 mm, a weight of 31 gr, a flow rate of 452 (ml/min), and a max continous power of 0.4 Watt. The motor is connected to a

[3] https://www.reynoldsam.com/product/sil-poxy/?pk_campaign=6615436476.

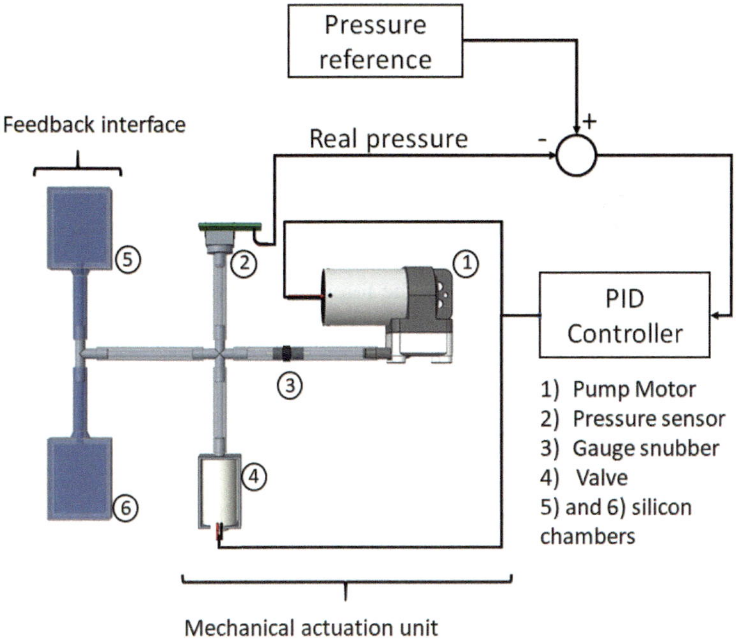

Fig. 7.3 Overview of the Soft Haptics pneumatic system

solenoid valve (4) KSV05A (rated Voltage of DC 3V and a rate current of 75 mA), and to an integrated silicone pressure sensor (2) MPXA4250AC6U (Pressure range of 20–250 kPa, sensitivity of 20 mV/kPa, an operation temperature from deg −40 to deg +125 C, and an accuracy of $\pm 1.5\ \% V_{FSS}$).

To reduce the pressure oscillation introduced by the pump we use a gauge snubber (3). It is worth noticing that the device presented in this work is only a proof of concept and the primary goal of our work was to explore if the WISH system could be a viable solution to deliver force feedback in prosthetic applications. Future work will consider an in-depth investigation of the technical aspects that are crucial for an everyday life usage of our system, including the characterization of the durability of the silicone chambers as well as of the actuation unit.

7.2.3 Electronic and Control System

We used a custom electronic board based on a Cypress Psoc (Programmable System-on-Chip) micro-controller, with RS485 communication protocol, to control the pump flow rate. The design of the electronic board is open source and more information can be found on the Natural Machine Motion Initiative website.[4]

The block diagram in Fig. 7.3 describes the closed loop control scheme. A PID controller modulates the difference between the reference pressure and the pressure readings from the sensor, and activates the pump with a PWM signal. We chose an operation range for the motor between 1 and 3.5 V, which corresponds to a duty cycle value of 20 and 80%. The choice of the lower limit was mandatory, because the pump does not activate with a duty cycle less than 20%. Whereas we fixed the upper limit to 80% so to reach a pressure of 25 kPa, which is the maximum value we fixed to obtain a normal force on the forearm equal to about 14–15 N (for more details see Sect. 7.3).

7.3 System Characterization

To characterize the system we performed: a physical characterization (Sect. 7.3.1), and a psychophysical characterization (see Sect. 7.3.2) of the pressure just noticeable difference (JND) at different upper limb locations. We also asked participants to undergo a subjective quantitative evaluation to understand the acceptability and effectiveness of the stimulation modality.

7.3.1 Physical Characterization

A 3-axis force sensor, the ATI Gamma 17 Mini,[5] was used to measure the normal force exerted by the WISH device. To be able to collect the data, a custom structure was built to adapt the sensor to the shape of the chamber. We 3D printed, in ABS, two different structures, one where the chamber was completely constrained inside a rigid location, one where the chamber had a rigid constraint on one side and a soft constraint, i.e. the forearm, on the other side (see Fig. 7.4).

First we tested the system using different PWM signals, to find a suitable operating range of pressure, which can be perceived on the forearm. We started using a duty cycle of 20%, and increased the value until 100%. We inflated the chamber for 5 s, and then deflated it opening the valve; each duty cycle value was repeated 10 times. Looking at the results obtained from the experiment (see Fig. 7.5), we decided to use a range of pressure between 15 kPa and 40 kPa, because in this range the force had a linear trend respect to different level of pressure (see Fig. 7.7).

We identified five different levels of pressure inside these limits, precisely $15, 21.25, 27.5, 33.75, 40$ kPa, equally spaced, and used the two custom structures for the characterization. The characterization inside the custom rigid structure

[4] https://www.naturalmachinemotioninitiative.com/copy-of-elettronica-v1.

[5] https://www.ati-ia.com/products/ft/ft_models.aspx?id=Mini45.

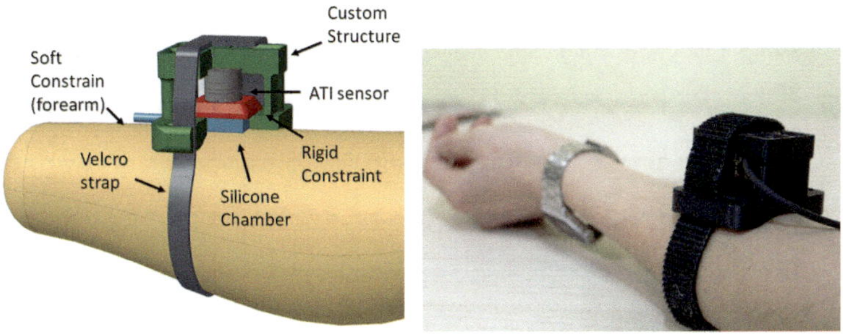

(a) (b)

Fig. 7.4 The overall architecture and components of the custom structure Fig. 7.4a. The structure worn by one subject, Fig. 7.4b

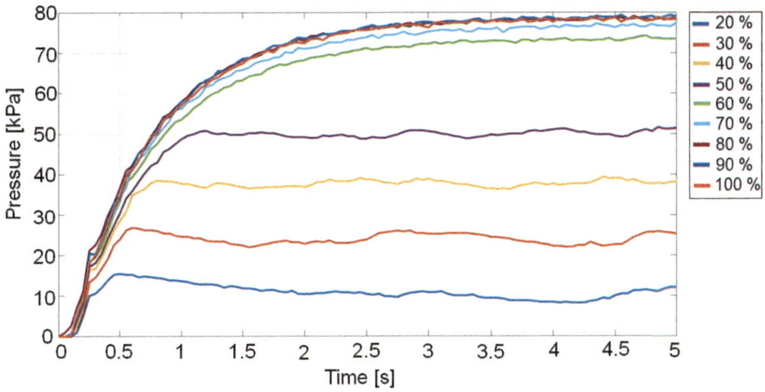

Fig. 7.5 The picture shows the different levels of pressure obtained with the duty cycle values. The legend, in the picture, indicates all the PWM values considered

was accomplished with the silicon chamber completely closed between two rigid constraints on both sides (see Fig. 7.4a, for the rigid constraint).

We inflated the chamber for 5 s and then deflated it; each pressure value was repeated for 5 times. We recorded the pressure and the force values corresponding for each inflate/deflate cycle. Figures 7.6 and 7.7 show the performance of the PID controller and the force elicited by the chamber on the force sensor.

We repeated the same protocol for the characterization with the soft constraint; in this case one side of the chamber was in contact with the rigid constraint, as before, and the other was directly in contact with the forearm of two subjects. The results obtained show that with the soft constraint, the system is capable to follow precisely the reference pressure (see Fig. 7.8). Also in this case the force profile had a linear trend respect to different level of pressure (see Fig. 7.9. This opens to the possibility to control the pressure to deliver controlled stimuli to the user's forearm.

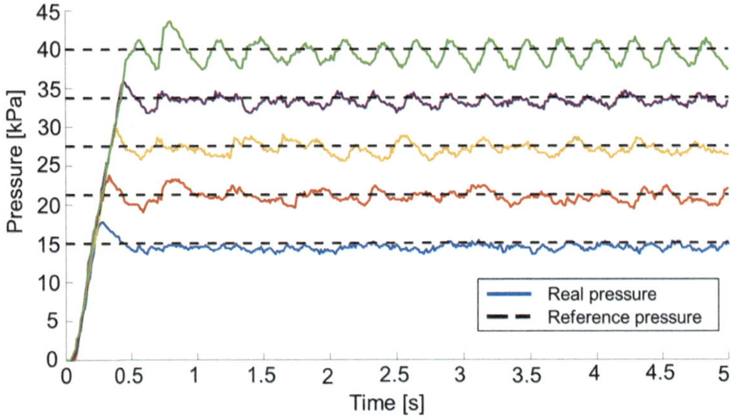

Fig. 7.6 Pressure evolution within the chamber when a rigid constraint was applied to limit its deformation. The dotted and the colored lines indicate the reference and the real measured pressure

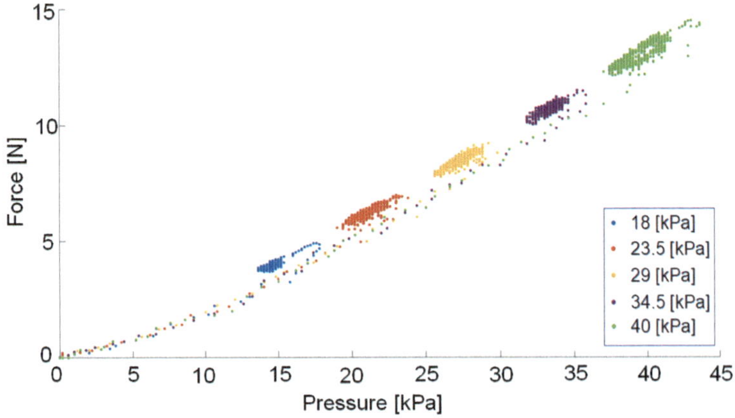

Fig. 7.7 Force evolution with the different pressure level tested during the characterization

7.3.2 Psychophysical Characterization

The goal of the Psychophysical characterization was to verify whether the five values of pressure, obtained by the physical characterization, can be clearly distinguished by the subjects.

We used the method of the constant stimuli to find the Just Noticeable Difference (JND), as defined in Gescheider (2013), i.e. the minimum amount of pressure that can elicit a different perception in users with respect to a reference stimulus.

We considered to place the chambers at two locations on the forearm, to ease the integration of the device in different sockets and for different residual limb conditions. In the first configuration, one chamber was placed on the glabrous skin and the other

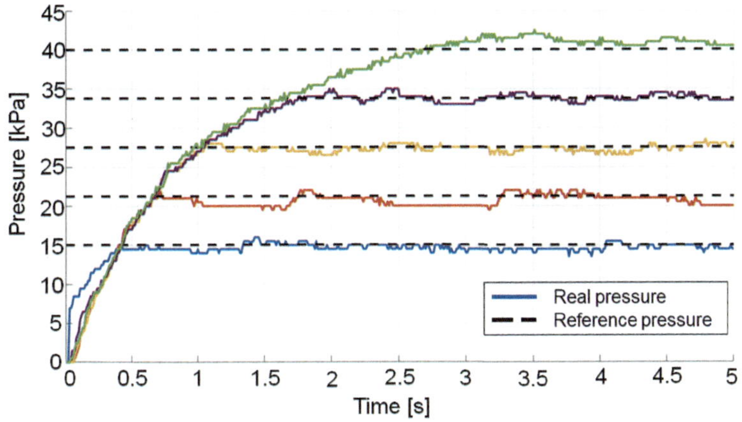

Fig. 7.8 System performance: the chamber is inside a soft constraint

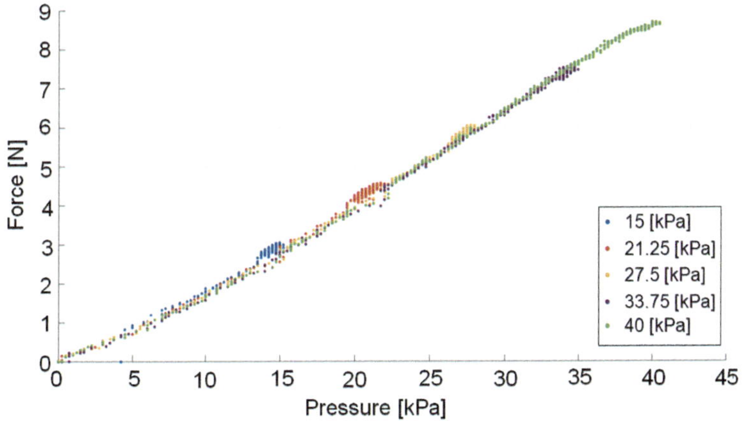

Fig. 7.9 Force evolution with the chamber inside a soft constraint

one on the hairy skin, while in the second configuration both the chambers were placed on the hairy skin (see Fig. 7.10a, b, respectively).

(1) *Participants*: Ten able-bodied participants (one female, mean age 29 and standard deviation of 2.9) took part to the experiment. None of them had experience in the use of soft pressure force feedback devices as sensory substitution techniques. Participants gave their informed consent before the experiment, and nobody had any physical or mental limitation which could have affected the experimental outcome. The experimental procedure was approved by the Ethical Committee Area Vasta Nord Ovest (CEAVNO) of the Tuscany region (Identification code n° 1072).

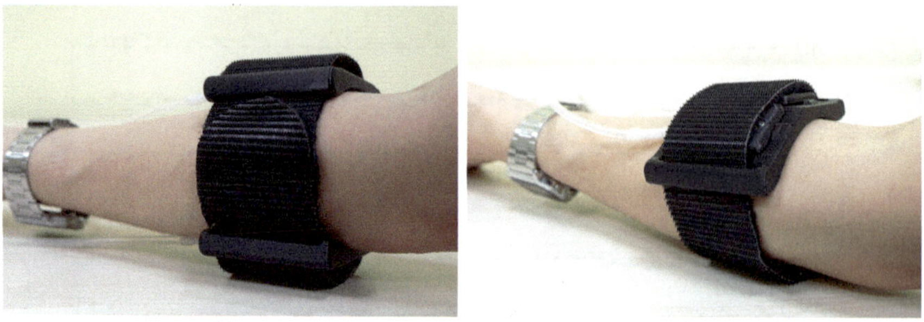

(a) (b)

Fig. 7.10 The WISH Device applied on a subject's arm **a** first configuration with one chamber on the glabrous skin and one on the hairy skin (bottom-up); **b** second configuration with both the chambers on the hairy skin only up

(2) *Stimulus and procedure*: The subjects were comfortably seated with their forearm resting on a desk, with the elbow at approximately 90°. They wore the WISH device fastened with a Velcro band on their forearm.
During the experiment, subjects wore headphones with pink noise to prevent the interference of any auditory cue; in addition they were required not to look at the device during the experiment. The role of the Velcro band is purely instrumental, since it aims at ensuring a proper contact of the silicone chambers with the skin of able-bodied participants. This contact is naturally achieved with prosthesis users thank to integration of the WISH device within the inner socket. Participants received paired stimuli, each stimulus was related to pressure value provided by the chamber on the skin, and were asked to indicate which stimulus in the pair was perceived as higher.
Each pair consisted of a reference stimulus (RS) and a comparison stimulus (CS), presented in random order. The pressure was equal to 27.5 kPa in the reference stimulus, and was pseudo-randomly chosen among five discrete and equally spaced values (corresponding to 15, 21.25, 27.5, 33.75, and 40 kPa Sect. 7.3.2) in the comparison stimulus.
A single trial consisted of: the first stimulus, an inter-stimulus interval (2 s), and the second stimulus followed by the subject's response. We considered two experimental blocks corresponding to the two locations on the forearm, and the order of execution of the two blocks were counterbalanced across the subjects. Each block consisted of 100 trials. The binary response of the participants was recorded.
At the end of the experiment, participants underwent through a subjective quantitative evaluation procedure based on a seven point Likert-scale survey, which consisted of questions about the system and the experimental tasks. Participants had to answer by assigning a score ranging from 1 *totally disagree* to 7 *totally agree*. This represents a common procedure to evaluate devices for assistive

Table 7.1 These statements, presented in random order, were rated by the able-bodied subjects using a 7-point Likert scale (1: Strongly disagree, 7: Strongly agree). Median and interquantile range across all individuals are reported

	Questions	Median	Iqr
Q1	It was easy to wear and use the WISH device	6.5	1
Q2	I was feeling uncomfortable usign the WISH	1	1
Q3	I was well-isolated from the external noises during the experiments	7	0
Q4	I was able to hear the sounds made by the actuator of the device	1	0
Q5	The stimuli provided by the cutaneous device allowed to discriminate different pressure levels	6.5	1
Q6	I felt hampered by the air pressure stimuli	1.5	1
Q7	I was able to see the device during the experiment	1	0
Q8	At the end of the experiment I felt tired	2	1
Q9	I prefer to receive the stimuli at two different body locations, i.e. glabrous and hairy skin	6	1
Q10	I prefer to receive the stimuli at the same body location, i.e. hairy skin	5	2

robotics and Human-Robot Interaction (Prattichizzo et al. 2013). The questionnaire considered the comfort and usability of the experimental setup (Q1, Q2, Q3, Q4 and Q7), the perceived performance (Q5, Q6, Q9 and Q10), and the engagement of the subject (Q8). Questions are presented in Table 7.1, and the results are shown in Fig. 7.12.

(3) *Data analysis* For the two experimental conditions, i.e. the body location for stimulus delivery, we positioned the chambers: (1) one in the glabrous skin and the other on the hairy skin—hereinafter referred to as condition *up-down-*, and (2) both on the hairy skin—hereinafter referred to as condition *only up*).

We used a Generalized Mixed Model GLMM to determine JND through all the participants (Moscatelli et al. 2012; Rossi et al. 2018). This model enabled cluster data analysis, in our case a cluster is the collection of repeated responses in several participants. GLMM is a hierarchical model that provides both predictions on the experimental effects, which are assumed to be systematic across participants, and an estimate of the variability between participants. The expected value of the response can be expressed as

$$\Phi^{-1}\left[P\left(Y_j = 1\right)\right] \sim \beta_0 + \beta_1 x_j \tag{7.1}$$

where $\left[P\left(Y_j = 1\right)\right]$ is the probability that, in trial j, the participant reported a larger stimulus in the comparison than in the reference stimulus, $\Phi^{-1}[\cdot]$ is the probit transformation of the response probability (i.e., the inverse function of the cumulative normal distribution), and x_j is the explanatory variable, i.e.

the pressure value provided to the chambers. β_0 and β_1 are the fixed effect parameters, i.e. the intercept and the slope of the linear function (linear predictor), which are the same for all the subjects.
For each experimental condition, we estimated the Just Noticeable Difference (JND), i.e. the amount of stimulus change to detect the just noticeable difference, the Point of Subjective Equality (PSE), i.e. the stimulus value yielding a response probability of 0.5, with the related 95% confidence intervals (CIs), using the bootstrap method (Moscatelli et al. 2012).

(4) *Results* The GLMM fitting to the data is illustrated in Fig. 7.11.
The JND was equal to 3.40 kPa (95% CIs: 2.72–4.15 kPa) for the up-down condition, and 3.38 kPa (95% CIs: 2.75–4.20 kPa) for the only up condition.
The PSE was equal to 27.51 kPa (95% CIs: 26.65–28.46 kPa) for the up-down condition and 27.47 kPa (95% CIs: 26.66–28.47 kPa) for the only up condition.
Results of the bootstrap method revealed that the 95% CI of the differences between JNDs and PSEs in the two conditions include zero. These results imply that these differences are not significant.

7.4 Integration Inside a Prosthesis

According to the results obtained from the experiment with able-bodied subjects, we integrated the system inside a inner socket positioning both the silicone chambers on the hairy skin (i.e. the only-up configuration of able-bodied subjects); Fig. 7.1 shows the WISH device inside the inner socket, while Fig. 7.13 shows a prosthetic user wearing the device.

The mechanical structure is composed of 3 main parts: the structural frame, the mechanical actuation and the feedback interface. The structural frame is composed of the SoftHand Pro (Godfrey et al. 2018), the socket and the inner socket inclusive of surface electromyographyc (sEMG) sensors (respectively parts $\{1, 2, 3, 5\}$ in Fig. 7.1). The mechanical part is composed of a battery pack (6), an electronic board (7), a solenoid valve (8), a motor pump (9), and a pressure sensor (10). The feedback interface is composed of the two silicone chambers (4) integrated in the upper part of the inner socket (see Fig. 7.15). All the components are connected with silicone pipes (11).

7.4.1 The Control Strategy

The block diagram in Fig. 7.14 describes the closed-loop control scheme of the SoftHand Pro with the human in the loop. The electromyographic activity from the Flexor Digitorum Superficialis (FDS) and the Extensor Digitorum Communis (EDC)

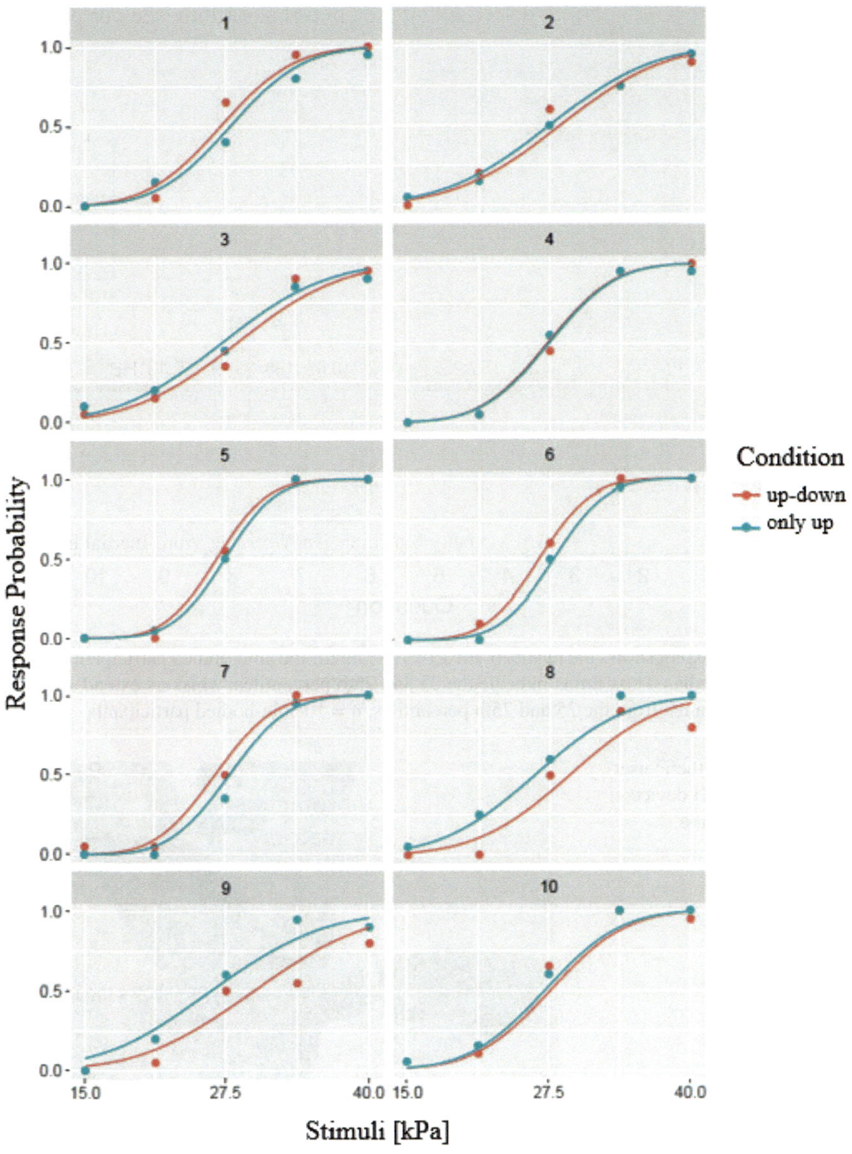

Fig. 7.11 GLMM fit for all the subjects. Each box represents raw data and model predictions for each single participant (labeled as 1–10) in the two condition

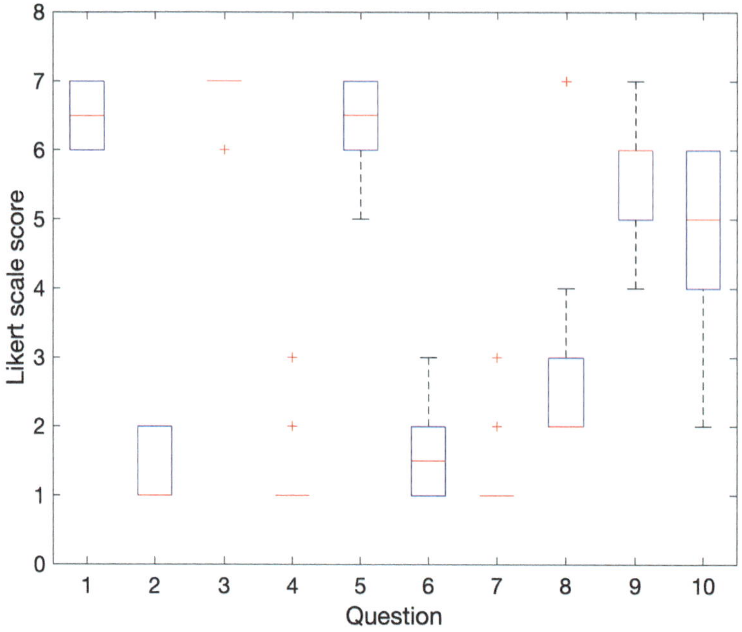

Fig. 7.12 The boxplot shows the results of the Likert Scale for the able-bodied participants. Center lines show the medians; box limits indicate the 25 and 75th percentiles; whiskers extend 1.5 times the interquartile range from the 25 and 75th percentiles. n = 10 able-bodied participants

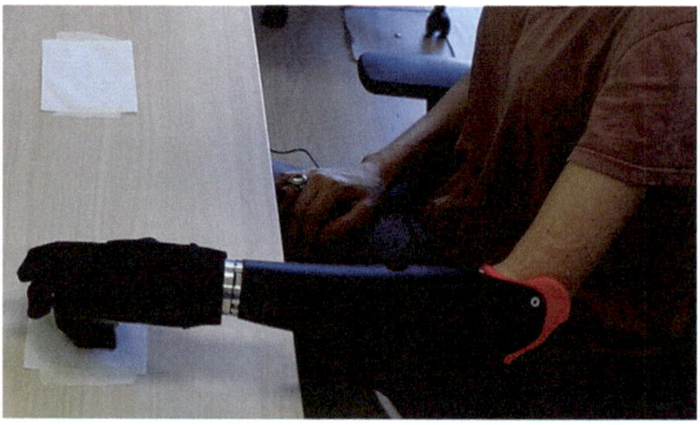

Fig. 7.13 A prosthetic user wearing the WISH device integrated inside the prosthesis

muscles is measured from the skin of the user's forearm using two Ottobock 13E200 sensors,[6] which directly provide an amplified, bandpass-filtered, and rectified version of the raw sEMG signals.

The reference pressure, p_{ref}, used to control the air pump and the valve is set to be proportional to an estimate of the grasp force exerted by the hand. To estimate the grasp force we used the difference between the estimated current and the real current

[6] Otto Bock HealthCare GmbH, http://www.ottobock.com/.

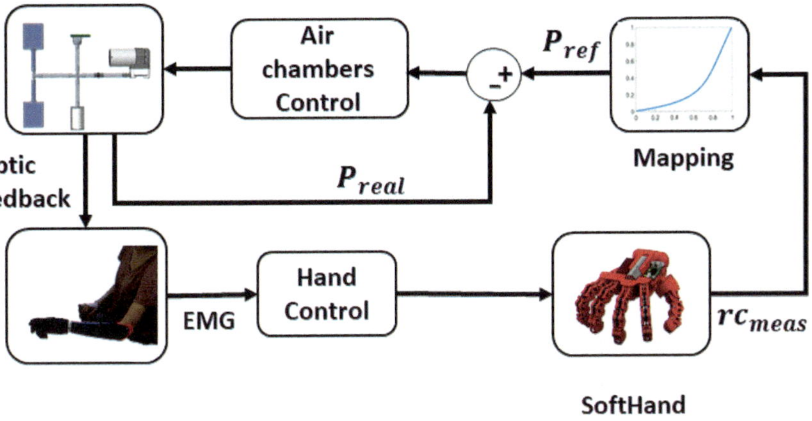

Fig. 7.14 Block diagram of the SoftHand Pro, controlled by the sEMG sensors, with the haptic feedback in the loop

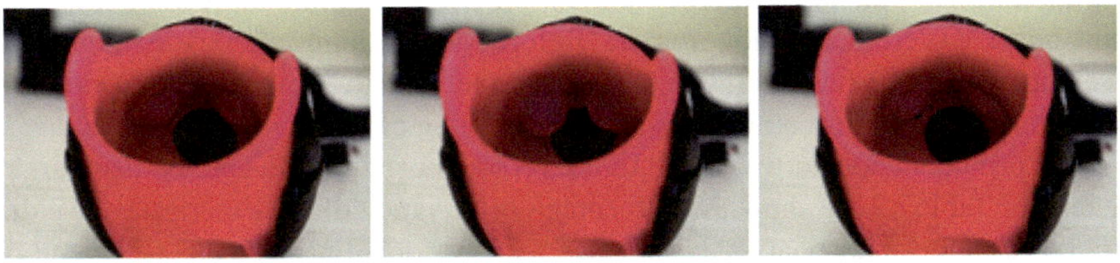

Fig. 7.15 Sequence of images that show the mode of operation of the chamber inside the inner socket

absorbed by the SoftHand Pro motor, called residual current, rc_{meas} (see Mura et al. 2018 for details).

The tactile stimulation performed by the WISH, directly depends on the value of the SoftHand Pro residual current; this information can be used to understand how much force the hand is applying during the grasp of an object. We establish two thresholds: if the residual current is over 500 mA, it is saturated to this value, the air pump is on, the valve is closed and the chambers start to inflate; if the residual current is equal or minus than zero, the value is set to zero, the pump is off, the valve is open and the chambers are deflated.

Mapping the information coming from the robotic hand into feedback device, means to associate each value of the hand with a corresponding pressure value of the device. It is not an on/off control but a real time modulation of the residual current value of the SoftHand into inflation/deflation of the silicone chambers. In our investigation we tried different types of mapping (Rossi et al. 2018), and the best mapping we found out was a parabolic mapping. We imposed three simple constraints: $rc_{meas} = 0$ corresponds to $p_{ref} = 0$, $rc_{meas} = rc_{max}$ corresponds to $p_{ref} = p_{max}$, being rc_{max} the maximum value of the residual current of the SoftHand, and p_{max}

the maximum pressure value inside the chambers; and $rc_{meas} = \frac{1}{2} rc_{max}$ was mapped to $p_{ref} = \frac{4}{5} p_{max}$. The reference position was:

$$p_{ref} = \frac{-30x^2 + 55 * 450x}{450^2} \quad (7.2)$$

where $x = rc_{meas} - 50.0$.

7.5 Pilot Experiment

We tested the integrated system with one amputee subject (female, age 39 with agenesis of left forearm).

The participant did not suffer from any cognitive impairment, that could have interfered with her ability to follow the instruction of the study. The subject wore her prosthesis, with the WISH integrated, on her left arm, and headphones with pink noise to prevent any auditory cues. Before starting doing other experiments, we accomplish the JND test also with the amputee subject, to verify if the perception of an amputees subject is comparable to the perception of an able-bodied subject.

The protocol adopted was equal to the protocol used for the able-bodied participants. In Table 7.2 we show the percentage of accuracy of the experiment. We did not perform any statistical analysis, since we have only one subject. At the end of the experiment the subject underwent through the subjective quantitative evaluation procedure, the Likert-scale survey (see Table 7.3).

We asked the subject to accomplish two different tasks: (1) a cylinder discrimination task, and (2) a virtual eggs task. In our hypothesis what the user really feels is the change in impedance of the chamber-skin-muscle system interpose between the rigid socket and the rigid bone, which can be probed by muscle activation. In addition, the diameter of the residual limb of the prosthesis user did not change noticeably when the muscle was relaxed or contracted (remaining approximately equal to 19.5 cm). For these reasons, we decided to not explicitly model force changes due to muscle size variation for this subject. This represent an important aspect to be considered in future investigations.

Table 7.2 The table reports the results of the comparison between the pair stimuli with the percentage of success

Comparison	I > ref	Success (%)
1–3	0	100
2–3	5	75
3–3	11	55
4–3	11	55
5–3	16	90

Table 7.3 These statements, presented in random order, were rated by the amputee subject using a 7-point Likert scale (1: Strongly disagree, 7: Strongly agree). Only the scores are reported

	Questions	Score
Q1	It was easy to wear and use the WISH device together with the SoftHand	7
Q2	I was feeling uncomfortable while using the WISH device together with the SoftHand	1
Q3	I was well-isolated from the external noises during the experiments	7
Q4	I was able to hear the sounds made by the actuator of the device	1
Q5	The stimuli provided by the cutaneous device allowed to discriminate different pressure levels	4
Q6	At the end of the experiment I felt tired	3
Q7	I felt hampered by the air pressure stimuli	1
Q8	I was able to see the device during the experiment	2
Q9	The sensation provided by the air pump on the forearm felt pleasant	6
Q10	The sensation provided by the air pump on the forearm felt unpleasant	2
Q11	I had the feeling of performing better while receiving force feedback by the air pressure chambers	6
Q12	I had the feeling of performing better when I was not receiving any force feedback by the air pressure chambers	1

These two experiments were designed to evaluate the effectiveness of the feedback provided by the WISH device when using the SoftHand Pro. At the end of both experiments the subject underwent the System Usability Scale (SUS) the most widely used standardized questionnaire for the assessment of perceived usability (Lewis 2018).

7.5.1 Cylinder Discrimination Task

For the cylinder-softness discrimination task, subject was comfortably seated in front of the table with the hand placed on the desk in a fixed position. The experiment lasted roughly one hour and consisted of two phases. First the training phase, which lasted approximately 20 min, the subject made manipulation tasks of different objects of various size, i.e. building pyramids with toy blocks, grasping and moving cylinders. Neither vision nor hearing were obstructed during this phase.

During the test phase, which lasted more or less 30 min, the subject was blindfolded and wore headphones with pink noise, to prevent any auditory cue. For the test, we selected three stimuli: the hand closed without object, and with a rigid and soft cylinder both, 60 mm in diameter. For convenience, the three cases were labelled as cylinder 0, 1 and 2, respectively.

In each trial, we provided participant with two stimuli, sequentially, and we asked her to report which one was the higher pressure. Each cylinder was positioned on the palm of the SoftHand Pro and the subject had to voluntarily close the hand to grasp it. Each pair was presented twice. The test was performed in two modalities, with and without feedback.

7.5.2 *Virtual Eggs Task*

We asked to the subject to accomplish a task similar to a Box and Block test: she had to move a virtual egg, from point A to point B (Fig. 7.16c, for ten times, without breaking it. This task was used in Engels et al. (2018) to assess the grasping efficiency.

The virtual egg is a box made by paper, dimension 50 × 50 mm containing a fragile fuse placed in the middle of the box, and a weight of 320 g (see Fig. 7.16a, b. The fragile fuse is a piece of spaghetto, fragile enough to be broken by an excessively tight pinch grasp. More information regarding the virtual eggs can be found on the Natural Machine Motion Initiative website.[7]

The subject stood in front of the table and repeated the experiment with and without the feedback; in both condition there were not limit of time to move the box. For each modality we recorded: (1) the time spend to move the fragile object from A to B (we started recording time when the subject touched the virtual egg and stopped when she released the box), (2) the number of regrip, and (3) the number of broken eggs. Table 7.4 shows the result of the experiment.

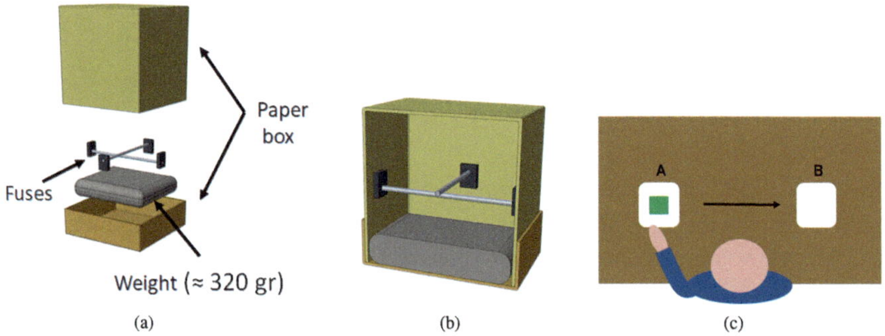

Fig. 7.16 Overview of the virtual egg Fig. 7.16a, b band the setup used for the second task

[7] https://www.naturalmachinemotioninitiative.com/fragile-object-benchmark.

Table 7.4 The table reports the time (in second) to move the fragile objects from point A to point B

Trial	Without WISH		With WISH	
	Time to move [s]	Broken eggs	Time to move [s]	Broken eggs
Task 1	6.42	1	5.70	0
Task 2	6.56	1	5.99	1
Task 3	6.58	0	8.08	0
Task 4	7.4	1	8.02	0
Task 5	5.92	1	5.98	1
Task 6	6.34	0	7.41	1
Task 7	4.10	1	5.29	0
Task 8	9.97	0	5.60	0
Task 11	6.24	0	5.81	0
Task 10	8.12	1	5.77	0

7.5.3 Results

The results of the cylinder-softness discrimination task show a major accuracy when the subject accomplishes the experiment with the device, respect to the condition without feedback. For the cylinder discrimination task, we obtained an accuracy of 100% for the modality with the feedback, and an accuracy of 75% for the modality without the feedback (see Table 7.2).

Table 7.4 shows the results of the fragile object. Although we cannot make any statistical claim, the number of broken eggs without feedback is 6, compared to the 3 with the feedback. Finally from the SUS questionnaire, emerged a positive rate with a total score of 70, which means a good device but it needs to be improved.

7.6 Discussion

The experimental outcome with the able-bodied and the limb loss participants show that the WISH device can be successfully used to convey the force feedback.

From the psychometric experiment and the Likert scale of able-bodied participants, emerges that there is no difference detecting the pressure in different parts of the forearm (questions Q9 and Q10 with mean values of 5.6 and 4.5, and a standard deviation 0.96 and 1.34). In general, subjects felt comfortable to wear and use the device (question Q1 and Q2 with a mean value of 6.5 and 1.3 and a standard deviation of 0.42 and 0.69). During all the experiments, subjects were well isolated from external noises (Q3, Q4 and Q7). The stimuli provided by the device were clear and allowed to discriminate different pressure levels (Q5).

Regarding the intuitiveness of the sensory substitution, no one of the participants felt hampered by the air pressure stimuli (Q6 with a mean of 1.3 and a standard deviation of 0.69). At the end of the experiment the participants were not tired (question Q8, mean value 1.6 and stardard deviation of 0.69).

The results obtained during the experiments with the limb loss participant are very promising. The survey outcome shows that the WISH is easy to wear (Q1) meaning that the device is not an obstacle during the wear phase of the prosthesis. Moreover, the subject was not hampered by the air pressure stimuli, and the sensations provided were pleasant (Q7, Q9 and Q10), so we can assume that the MM paradigms is successfully satisfied. The sensation provided to the residual limb is helpful and clear to understand (Q12). The subject was able to recognize different level of pressure during passive and active task (i.e. the cylinder-softness and virtual egg task respectively).

During the active task, the subject, using the device, was able to modulate the force in order to grasp the box without broke the fuse inside it. The score of SUS questionnaire underline a positive outcome.

7.7 Conclusions

We presented the WISH device, a pneumatic force feedback used to convey information on the grasp force of a prosthetic hand that can be integrated in the socket.

The psychometrics analysis shows that there is no difference in chamber placement for the two forearm locations that were tested. Experimental outcomes with the prosthetic user reveal a good level of accuracy for the softness discrimination task, and a good control of the modulation of the grasping force during manipulation tasks.

In conlusion, the WISH device seems to be a viable solution to provide a force feedback completely integrated inside a socket. Future work will focus on further testing with real prosthesis users, including a deeper investigation of the effect of muscle size variations with contraction. While a 1D force feedback output is sufficient for the purpose of the current work (to deliver force-related information based on an indirect estimation of the overall grasping force of the SoftHand), in the future we will consider other prosthetic hands (including versions of the SoftHand with additional degrees of actuations Della et al. 2017), where multi-dimensional force feedback could be useful.

A possible direction include the use of more chambers, as proposed in Wijk et al. (2020), controlled by multiple air pumps. Another way could be to combine the air pressure with particle jamming, as explored by Stanley and Okamura (2015), to model the output feedback according to different directions.

Acknowledgements The authors would like to thank Nicoletta Colella, Cristiano Petrocelli and Mattia Poggiani for their useful help in the characterization process and system design.

References

Antfolk C, D'alonzo M, Rosén B, Lundborg G, Sebelius F, Cipriani C (2013) Sensory feedback in upper limb prosthetics. Expert Rev Med Dev 10(1):45–54

Antfolk C, Björkman A, Frank SO, Sebelius F, Lundborg G, Rosen B (2012) Sensory feedback from a prosthetic hand based on air-mediated pressure from the hand to the forearm skin. J Rehabil Med Official J UEMS Euro Board Phys Rehabil Med 44(8):702–707

Biddiss E, Beaton D, Chau T (2007) Consumer design priorities for upper limb prosthetics. Disabil Rehabil Assistive Technol 2(6):346–357

Clark JP, Kim SY, O'Malley MK (2018) The rice haptic rocker: Comparing longitudinal and lateral upper-limb skin stretch perception. In: Haptics: science, technology, and applications: 11th international conference, EuroHaptics 2018, Pisa, Italy, June 13-16, 2018, Proceedings, Part II 11. Springer International Publishing, pp 125–134

Clemente F, D'Alonzo M, Controzzi M, Edin BB, Cipriani C (2015) Non-invasive, temporally discrete feedback of object contact and release improves grasp control of closed-loop myoelectric transradial prostheses. IEEE Trans Neural Syst Rehabil Eng 24(12):1314–1322

Culbertson H, Schorr SB, Okamura AM (2018) Haptics: The present and future of artificial touch sensation. Ann Rev Control Robot Auton Syst 1:385–409

Della Santina C, Piazza C, Gasparri GM, Bonilla M, Catalano MG, Grioli G, Garabini M, Bicchi A (2017) The quest for natural machine motion: an open platform to fast-prototyping articulated soft robots. IEEE Robot Autom Mag 24(1):48–56

Engels LF, Cappello L, Cipriani C (2018) Digital extensions with bi-axial fingertip sensors for supplementary tactile feedback studies. In: 2018 7th IEEE international conference on biomedical robotics and biomechatronics (Biorob). IEEE, pp 1199–1204

Fani S, Di Blasio K, Bianchi M, Catalano MG, Grioli G, Bicchi A (2019) Relaying the high-frequency contents of tactile feedback to robotic prosthesis users: design, filtering, implementation, and validation. IEEE Robot Autom Lett 4(2):926–933

Fraser C, Wing AW (1981) A case study of reaching by a user of a manually-operated artificial hand. Prosthet Orthot Int 5(3):151–156

Gescheider GA (2013) Psychophysics: the fundamentals. Psychology Press

Godfrey SB, Zhao KD, Theuer A, Catalano MG, Bianchi M, Breighner R, Bhaskaran D, Lennon R, Grioli G, Santello M, Bicchi A, Andrews K (2018) The SoftHand Pro: functional evaluation of a novel, flexible, and robust myoelectric prosthesis. PloS One 13(10):e0205653

Gurari N, Kuchenbecker KJ, Okamura AM (2012) Perception of springs with visual and proprioceptive motion cues: implications for prosthetics. IEEE Trans Human-Mach Syst 43(1):102–114

Huaroto JJ, Suarez E, Krebs HI, Marasco PD, Vela EA (2018) A soft pneumatic actuator as a haptic wearable device for upper limb amputees: Toward a soft robotic liner. IEEE Robot Autom Lett 4(1):17–24

Katz D, Krueger LE (2013) The world of touch. Psychology Press

Kim K, & Colgate JE (2012) Haptic feedback enhances grip force control of sEMG-controlled prosthetic hands in targeted reinnervation amputees. IEEE Trans Neural Syst Rehabil Eng 20(6):798–805

Lewis JR (2018) The system usability scale: past, present, and future. Int J Human-Comput Interact 34(7):577–590

Moscatelli A, Mezzetti M, Lacquaniti F (2012) Modeling psychophysical data at the population-level: the generalized linear mixed model. J Vis 12(11):26

Mura D, Barbarossa M, Dinuzzi G, Grioli G, Caiti A, Catalano MG (2018) A soft modular end effector for underwater manipulation: a gentle, adaptable grasp for the ocean depths. IEEE Robot Autom Mag 25(4):45–56

Oddo CM, Raspopovic S, Artoni F, Mazzoni A, Spigler G, Petrini F, Giambattistelli F, Vecchio F, Miraglia F, Zollo L, Di Pino G, Camboni D, Carrozza MC, Guglielmelli E, Rossini PM, Faraguna U, Micera S (2016) Intraneural stimulation elicits discrimination of textural features by artificial fingertip in intact and amputee humans. elife 5:e09148

Piazza C, Grioli G, Catalano MG, Bicchi AJAROC (2019) A century of robotic hands. Ann Rev Control Robot Auton Syst 2:1–32
Prattichizzo D, Chinello F, Pacchierotti C, Malvezzi M (2013) Towards wearability in fingertip haptics: a 3-dof wearable device for cutaneous force feedback. IEEE Trans Haptics 6(4):506–516
Rossi M, Bianchi M, Battaglia E, Catalano MG, Bicchi A (2018) HapPro: a wearable haptic device for proprioceptive feedback. IEEE Trans Biomed Eng 66(1):138–149
Stanley AA, Kuchenbecker KJ (2012) Evaluation of tactile feedback methods for wrist rotation guidance. IEEE Trans Haptics 5(3):240–251
Stanley AA, Okamura AM (2015) Controllable surface haptics via particle jamming and pneumatics. IEEE Trans Haptics 8(1):20–30
Svensson P, Wijk U, Björkman A, Antfolk C (2017) A review of invasive and non-invasive sensory feedback in upper limb prostheses. Expert Rev Med Dev 14(6):439–447
Wijk U, Carlsson IK, Antfolk C, Björkman A, Rosén B (2020) Sensory feedback in hand prostheses: a prospective study of everyday use. Front Neurosci 14:505038

Chapter 8
Toward Modality Matching: The Wearable Vibro-Inertial Bionic Enhancement System (VIBES)

Abstract In this chapter, we focus on the essential paradigm of modality matching for haptic feedback. Modality matching refers to the natural association between a sensation and the corresponding stimulus. To achieve this, we designed a vibrotactile device fully integrated into a prosthetic socket, capable of transmitting high-frequency contact information such as surface roughness and initial contact from interactions with external objects. The device features two compact planar vibrotactile actuators that are in direct contact with the user's skin, providing tactile cues. These cues correspond to the acceleration profiles recorded by two IMUs placed on the distal phalanx of a soft under-actuated robotic prosthesis (SoftHand Pro). We characterised the system from a psychophysical perspective with fifteen able-bodied participants, measuring their Just Noticeable Difference (JND) in discriminating vibrotactile cues delivered to the index finger during the exploration of various sandpapers. Furthermore, we conducted a pilot experiment with one user of the SoftHand Pro prosthesis, implementing a task called Active Texture Identification to determine if our feedback could enhance the user's ability to discriminate between different textures. The results indicate that the device effectively conveys contact and texture cues, which users can readily detect and distinguish.

8.1 Introduction

Sensory feedback is widely recognized as a priority for prosthetic users (Cordella et al. 2016; Biddiss et al. 2007), given its fundamental role in our cognitive processes, emotional state, and behaviour. Extensive research efforts have been directed towards methods to reproduce haptic feedback after the loss or absence of a hand to relieve prosthetic users from the need for constant visual attention (Motamedi et al. 2016). However, sensorized prostheses with feedback systems are currently only available through a few commercial options, as in Akhtar et al. (2020). Furthermore, developing a non-invasive sensorized prosthesis demands careful consideration of which haptic information transmit to the user and how, keeping the entire system intuitive, wearable, and integrated.

F. Barontini, *Wearable Haptic Devices for Realistic Scenario Applications*, Springer Series on Touch and Haptic Systems,
https://doi.org/10.1007/978-3-031-70539-7_8

Non-invasive feedback can be conveyed through various methods, with the most prevalent ones being electrotactile, mechano-tactile (force stimuli), and vibrotactile stimulation. Electrotactile stimulation is a non-invasive feedback solution that involves the application of electrical currents to the skin, thereby activating tactile nerve fibers (Bensmaia et al. 2023). The electrotactile electrodes are lightweight and can be easily integrated into the prosthesis. However, they can elicit unnatural sensations often considered unpleasant, high electrical voltage is required, and the presence of electrical fields can interfere with the myoelectric measurements necessary for prosthesis control (See et al. 2022; Bensmaia et al. 2023). On the other hand, vibrotactile stimulation has emerged as one of the most extensively investigated techniques for its ease in tactile signal modulation and for vibrotactile actuators' compact size, affordability, and availability (Choi and Kuchenbecker 2012; Thomas et al. 2019). Various methods have been suggested for converting a prosthetic hand's contact cue with an object and grip force into vibrotactile actuation, such as in Raveh et al. (2018), Markovic et al. (2018). However, despite the positive outcomes in terms of performance time in manual dexterity tests, grip, and manipulation accuracy, force vibrotactile feedback systems are not always integrated within the prosthetic device. Indeed, the vibrotactile actuators are usually placed with elastic bands on the arm or wrist of able-body subjects or on prosthetic users' contralateral arm and residual limb (Raveh et al. 2018; Markovic et al. 2018). When using a prosthetic device in daily living, having a separate feedback system, such as an additional device, can be uncomfortable and inconvenient due to lack of integration.

According to Kim et al. (2012) an optimal haptic feedback system should ensure the somatotopic matching (SM) and the modality matching (MM) paradigms. The somatotopic-matched feedback creates a natural signal comparable to the original sensation for the same body part. Antfolk et al. (2012) have demonstrated that sensory perception due to contact stimuli at the stump level can be processed by the central nervous system similar to that elicited by stimuli at the finger level. On the other hand, the modality-matched feedback provides a stimulus similar to the original sensation, i.e. pressure stimuli as a force cue. Vibrotactile stimuli are naturally associated with high-frequency information arising from surface contact, such as texture, roughness and first-contact cues (Shao 2022). Therefore, under the MM paradigm, the use of vibrotactile stimulation should convey acceleration-mediated contact cues related to the characteristics of objects being touched and facilitate their manipulation (ten Donkelaar et al. 2020). Nevertheless, methods to reproduce cutaneous cues of the surfaces as texture for prosthetic users are less investigated than solutions to communicate contact and grip force of prosthetic limbs (Balamurugan et al. 2019). Furthermore, to the best of the authors' knowledge, commercial embedded vibrotactile prosthetic feedback solutions for texture and contact artificial perception are still not currently available. Methods to deliver texture and contact cues include the Rehand II (Sakuma et al. 2020). The device is endowed with a skin vibration sensor and three vibrotactile actuators. The system undergoes testing with eight able-bodied adults, yielding favourable outcomes in relation to contact and operation recognition tasks. One of the very few integrated and commercialized vibrotactile feedback is presented in Akhtar et al. (2020). The PSYONIC Ability Hand is endowed with a single vibrator motor. Pressure sensors are placed on all five digits and mapped into the motor. While the PSYONIC Ability Hand is an advanced and sensorized prosthesis that conforms to the somatotopic paradigm, matching pressure cues with motor vibration could potentially increase the cognitive load needed to interpret haptic feedback. In our pursuit to advance tactile feedback technology with an intuitive, wearable, and embedded solution, we present a novel Vibro-Inertial Bionic Enhancement System (VIBES) (Fig. 8.1) in this work integrated with a SoftHand Pro (SHP) (Godfrey et al. 2018) prosthetic hand. In our previous work (Fani et al. 2019), we tested a MM vibrotactile system that recorded accelerations from the fingers of an artificial hand and reproduced them on a prosthetic user's skin through voice-coil actuators. In this work, the feedback system comprises two Inertial Measurement Units (IMUs) on the prosthetic thumb and index fingernails as sensors and two integrated vibrotactile actuators, efficiently transmitting texture and contact cues to specific stump sites to achieve both MM and SM. The presence of intrinsic somatosensory feedback generated by artificial body parts has been shown to be crucial for accurate motor commands (Amoruso et al. 2022). Thus, incorporating vibrotactile actuators in direct contact with the stump not only complies with the MM paradigm but also addresses the absence of intrinsic somatosensory feedback caused by the damping elements present in the SHP.

8.2 The VIBES System

As previously mentioned, the system is fully integrated inside the inner socket, with two vibrotactile actuators. Figure 8.1 provides an overview of the VIBES system. The structure consists of three main components: the structural frame, the sensor and electronics part, and the feedback interface. The structural frame includes the SoftHand Pro (SHP), the socket, and the inner socket, which contains the surface electromyographic (sEMG) sensors (parts 1, 2, 3, and 5). The sensor and electronics part comprises a battery pack (6), an electronic board (7), and two IMUs (8) positioned on the fingers of the hand. The feedback interface integrates two planar vibrotactile actuators (4) within the inner socket.

The vibrotactile actuators used in this study are Haptuator Planar (HP) actuators by TactileLabs[1] (See Fig. 8.2 on the right). The HP is made of a coil enclosing a permanent magnet. It has a 50–500 Hz high bandwidth which is within the range of tactile sensitivity. The transmission of stimuli is normal to the skin to reduce the size of the system with respect to tangential stimuli transmission. Its lightweight (1.8 gr), compact shape (12 x 12 x 6 mm), and soft surface designed to be directly in contact with the skin make the HP suitable for integration into prosthetic devices.

[1] Haptuator Planar by TactileLabs, [Online], Available: http://www.tactilelabs.com/products/haptics/haptuator-planar/

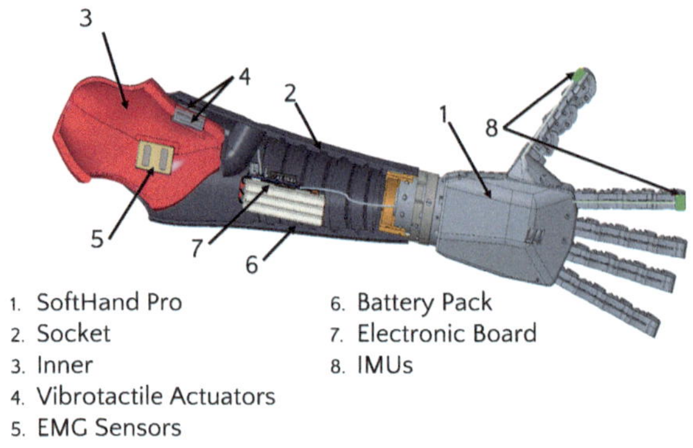

Fig. 8.1 Overview of the main components of the VIBES: Vibro-inertial bionic enhancement system

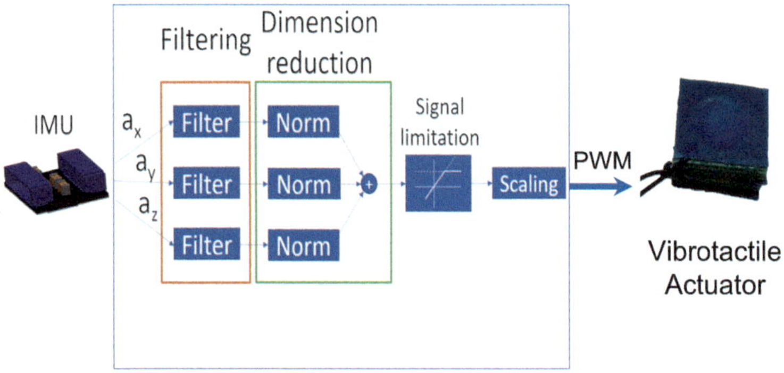

Fig. 8.2 Control strategy diagram of the vibrotactile system. The acceleration signals acquired from the IMU (left) undergo a signal elaboration process (filtering, dimensional reduction, and other minor elaboration steps). The processed signals are encoded into a PWM signal and transmitted to the actuator (right)

The control strategy of the VIBES is shown in Fig. 8.2. The acceleration signals $a = (a_x, a_y, a_z)$ recorded by each IMU are filtered, dimensional reduced and limited as in Fani et al. (2019). The filtering stage aims at removing unwanted artefacts caused by accelerometer readings during free-hand motion and vibrations from the prosthesis actuation system. Then, a scaling factor is applied to map the acceleration signals to a value of PWM to activate the actuator, considering the actuators' current limits. To achieve the somatotopic paradigm, the index IMU signal is mapped to the left actuator and the thumb IMU signal to the right actuator. The control works in real-time, and electronic boards (Santina et al. 2017) are used to record signals and drive the actuators. For further details about the control algorithm, see our previous work in Fani et al. (2019).

8.3 System Characterization

To test the effectiveness of the actuators in conveying reliable vibrotactile cues, we perform a physical and psychophysical characterization of the system.

8.3.1 *Physical Characterization*

The HP performance in rendering tactile input signals is tested. During tactile exploration, we save acceleration signals from an IMU on the robotic hand. Then, we measure the HP rendering by placing another IMU directly on the actuator. Figure 8.3 reports the comparison between an acceleration signal s and its rendering r. In this phase, no filtering techniques or signal reduction are applied. Based on the results obtained, which show the system's attempt to replicate the trend of the reference input signal, we set the control algorithm in Sect. 8.2.

8.3.2 *Psychophysical Characterization*

The experiment evaluates the capacity of human participants to discriminate different levels of roughness rendered by the actuators, corresponding to different sandpaper signals, in terms of Just Noticeable Difference (JND). JND refers to the minimum change in a stimulus required to produce a detectable difference in sensation (Jones and Tan 2012). We use the method of constant stimuli (Jones and Tan 2012), a common psychophysical technique, to calculate JND by presenting two stimuli and asking the participant to judge which one is greater. The experiment includes a *recording session* and a *user session*. In the *recording session*, the external tactile stimuli are recorded from the index finger of the SoftHand Pro and then saved. The index finger is chosen as the most commonly used finger for tactile exploration. In this stage, no filtering techniques are used. In the *users' session*, the tactile stimuli

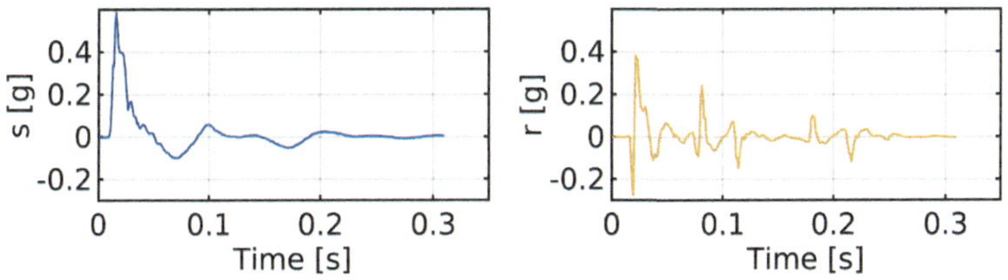

Fig. 8.3 Comparison between an acceleration signal s (left) and its rendering by the HP actuator r (right)

are reproduced on the users' fingers with the HP. All the experimental procedures are approved by the Committee on Bioethics of the University of Pisa-Review N. 30/2020.

8.3.2.1 Participants

Fifteen able-bodied right-handed participants (7 females and 8 males, age mean$\pm$SD: 26,6$\pm$2,05) are enrolled and asked to give their informed consent to participate in the experiments. Participants have no disorder that could affect the experimental outcome.

8.3.2.2 Recording Session

Five discrete stimuli are recorded in the *recording session*. The signal comes from sliding the SoftHand Pro index finger on different sandpapers. The set of sandpapers comes from the same manufacturer, RS,[2] and grit numbers and particle sizes shown are according to a preliminary study is done to determine the stimuli range. The smoothest and the roughest sandpapers detectable by the actuator are identified. The average grit sizes vary from 18 μm (grit number P1000) to 264 μm (grit number P60). To select the five stimuli, following (Libouton et al. 2010) procedure, we decide to use as reference stimulus P120 (average particle size of 127 μm), about halfway between the smoothest and the roughest stimuli. The other two stimuli are chosen to keep the stimuli equally spaced (in terms of micrometres), considering the existing types of sandpapers. Furthermore, according to the method of constant stimuli (Jones and Tan 2012), the extreme stimuli are chosen to be easily distinguishable from the reference stimulus, while the other stimuli are intentionally made challenging to differentiate. The selected sandpaper average grit sizes and grit numbers are shown in Table 8.1. Hence, five different sandpapers of 28 $\times$ 23 cm are used to record the signals.

Table 8.1 The five sandpapers selected: FEPA P-GRADE and average grit size

	FEPA P-GRADE	Average grit size (μm)
Smooth	P1000	18
	P220	65
Reference	P120	127
	P80	195
Rough	P60	264

[2] RS, [Online], Available: https://it.rs-online.com/web/.

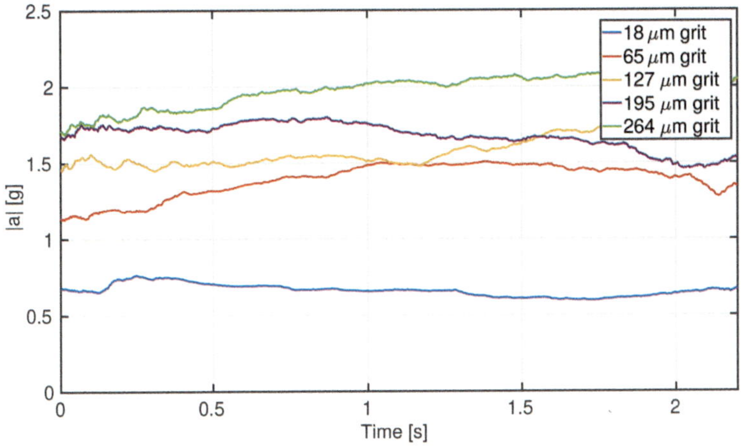

Fig. 8.4 Sandpaper acceleration signals: acceleration module |a| recorded for each sandpaper type

A custom C++ software is developed to register, cut, and save the accelerations coming from the IMU on the index fingertip. The three components of the acceleration $a = (a_x, a_y, a_z)$ are recorded from the IMU. The experimenter tries to keep velocity and pushing force as constant as possible by keeping fixed the time window to complete the recordings and taking the SHP at the same height above the paper. As reported in Boundy-Singer et al. (2017), passive texture presentation at varying scanning speeds had minimal effect on perception. Moreover, the perception of texture is also invariant to changes in contact force (Saal et al. 2018). Thus, velocity magnitude and pushing force should not influence the experiment results. Finally, two signals per sandpaper type are saved. Figure 8.4 shows the module of five different sandpaper signals recorded. A moving average filter is applied to improve the reading of the graph and facilitate the comparison. Each acceleration signal is mapped to a value of PWM to activate the HP, following the control strategy in Sect. 8.2.

8.3.2.3 Experimental Protocol

During the experimental session, participants are comfortably seated on an office chair, with their arm (elbow and forearm) placed on a desk. A custom 3D printed case is designed to enable the participants to use the HP comfortably. Participants are asked to keep their right index finger on the HP. The participants are isolated by wearing goggles with obscured lenses and headphones playing white noise. The experimenter verifies that the index finger lays on the case without putting pressure on it and without changing the finger position. A custom C++ software is developed to automatically manage the experiment following the requirements of the psychophysical method chosen.

The test involves presenting a reference and a comparison stimulus to the participant, with an audible signal announcing each pair. The reference stimulus is always P120, and the participant must indicate which stimulus felt rougher. There is no familiarization phase, and stimuli are not repeated. The comparison signal is randomly selected from the five pre-registered signals and presented to the subject in a pseudo-randomized order. Moreover, both the reference and the comparison signals are randomly chosen from the two stimuli saved for each sandpaper type. Each experiment consisted of 100 trials, 20 trials per stimulus level (corresponding to the five sandpaper grit).

8.3.2.4 Data Analysis

By means of a Generalized Linear Mixed Model (GLMM), we test the variability of the perceived roughness of the 15 subjects. GLMM account for the effect of the experimental variables of interest and the heterogeneity between clusters through fixed- and random-effect parameters, respectively (Moscatelli et al. 2012). We used the following model:

$$\phi^{-1}[P(Y_j = 1)] = \beta_0 + \beta_1 * x_j, \tag{8.1}$$

where $\phi^{-1}[\cdot]$ is the probit link function, which is due to dependent dichotomous variable, $P(Y_j = 1)$ is the probability of perceiving the comparison stimulus as rougher than the reference in trial j, β_0 and β_1 are the fixed effect parameters, i.e. the intercept and the slope of the linear function (linear predictor), which are the same for all the subjects. The explanatory variable is x_j, i.e. the sandpaper stimuli. Next, we estimated the JND and the Point of Subjective Equality (PSE), which is the stimulus value yielding a response probability of 0.5, with the related 95% confidence intervals (CIs), using the bootstrap method (Moscatelli et al. 2012). JND and PSE were computed as in Moscatelli et al. (2012):

$$JND = \frac{1}{\beta_1}, \tag{8.2}$$

$$PSE = -\frac{\beta_0}{\beta_1}. \tag{8.3}$$

8.3.3 *Results*

The GLMM fitting to the data is illustrated in Fig. 8.5. We evaluated the capacity of participants to discriminate roughness stimuli rendered by the HP actuator. We model 12 subjects' data as in (8.1). Three outliers were identified, exhibiting flat

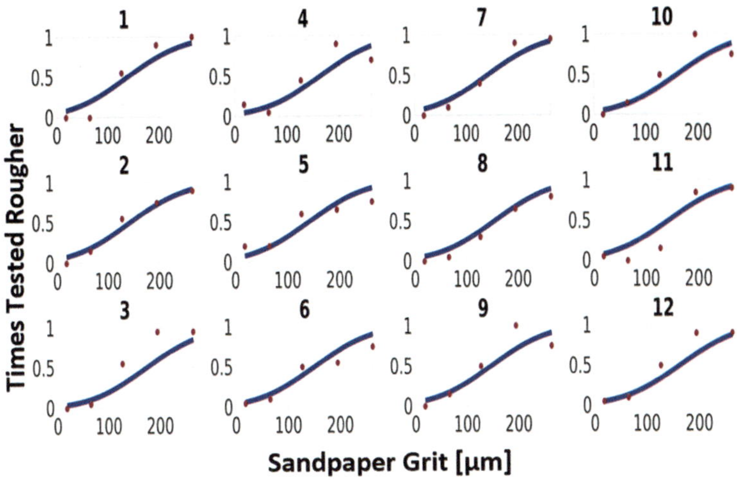

Fig. 8.5 GLMM fit for the 12 subjects. Raw data and model predictions for each participant labeled as 1–12

curves and indicating a low sensitivity to the texture (i.e. the subjects did not perceive significant distinctions among the five stimuli). The JND was equal to 87.30 μm (95% CIs: 79.69–96.5251 μm). The PSE was equal to 151.96 μm (95% CIs: 122.77–187.32 μm).

8.4 Pilot Experiment with a Prosthetic User

We tested the VIBES with one prosthetic user (age 43, female) affected by agenesis of the left forearm. The subject does not suffer from any cognitive impairment that could have interfered with her ability to follow the instructions of the study. The participant usually wears a cosmetic prosthesis, although she has experience with myoelectric ones. We integrated the VIBES inside an inner socket. Due to spatial constraints within the inner socket resulting from the presence of EMG sensors, we positioned both HP actuators on the hairy skin. This arrangement was chosen to effectively address the limited space available and ensure the appropriate placement of the actuators within the designated configuration. Figure 8.6 shows the VIBES inside the inner socket. Experiments aim to assess the efficacy of the feedback provided by the VIBES device with the SoftHand Pro in isolated conditions to reflect low-light situations and allow users to focus on the feedback (Raveh et al. 2018). The tests assess roughness discrimination and identification abilities by transmitting first-contact cues to the subject. For the SHP, the same control strategy as in Barontini et al. (2021) is used. At the end of the experiments, the subject completes the System Usability Scale (SUS) to evaluate the system (Lewis 2018).

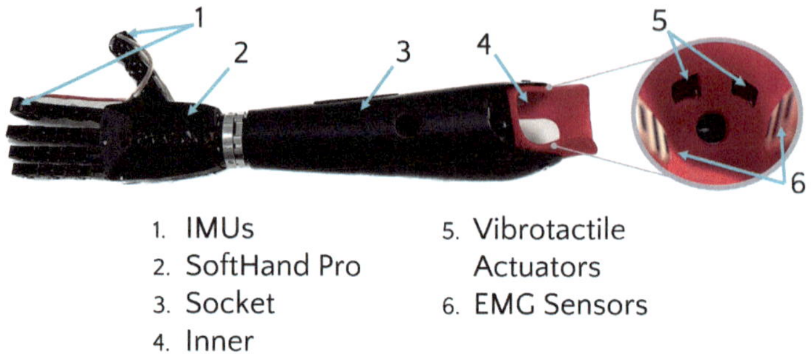

Fig. 8.6 The VIBES integrated inside the prosthesis. A detailed view (right) shows the inner part of the socket with vibrotactile actuators and EMG sensors

All the experimental procedures are approved by the Committee on Bioethics of the University of Pisa-Review N. 30/2020.

8.4.1 Psychophysical Characterization Experiment

A psychophysical characterization of a prosthetic user is performed to assess whether the subject's perception is similar to that of able-bodied individuals. The experimental protocol for the prosthetic user is the same as the one used for the able-bodied participants in Sect. 8.3.2. Thus, only the left HP is activated.

8.4.2 Active Texture Identification Experiment

For active texture identification, the subject is seated on a chair in front of a table. First, during 10 min familiarization phase, the participant is allowed to freely explore a P150 sandpaper (different from the one used in the subsequent experiment) fixed on the table while the VIBES is turned on. No isolation is provided in this phase. Then, during the experiment, the subject is isolated with white noise and obscured lenses. Figure 8.7 shows the setup of the experiment. The same sandpapers used in Sect. 8.3.1 are used in this task. The set of five sandpapers in descending roughness order, numbered from 1 (P60) to 5 (P1000), is placed on the right side of the table, which the subject explores using the right hand. On each trial, one sandpaper is presented to the subject under the left prosthetic hand for exploration using the prosthetic device. The participant is asked to identify the matching sandpaper from the five options on the right side. The task is performed in two modalities, with and without the VIBES feedback, and each sandpaper is randomly presented five times

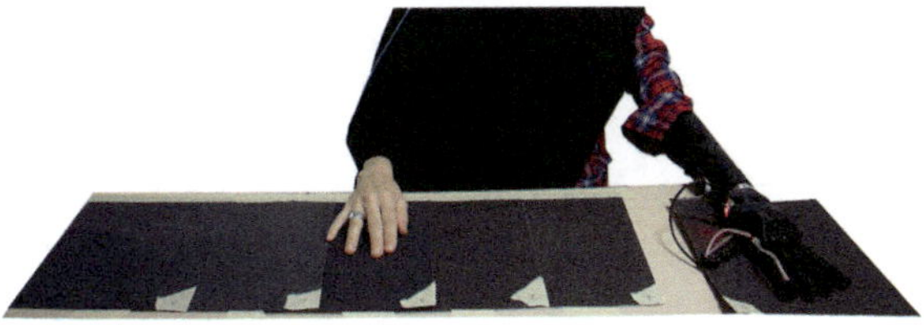

Fig. 8.7 The prosthetic user during the active texture identification experiment

Table 8.2 Results of the comparison between the pair stimuli with the percentage of success

Comparison	I >Ref	% Success
1–3	0	100
2–3	10	50
3–3	7	65
4–3	19	95
5–3	20	100

for a total of 25 identifications. Confusion matrices and accuracy metrics are used to analyze subject performance (Grandini et al. 2020).

8.4.3 Results

Since only one subject was involved, no statistical analysis was performed for the psychophysical characterization experiment. The accuracy results are shown in Table 8.2; from a probit fit on the data, the JND was 44.07 μm.

The results of the Active Texture Identification experiments are shown in Fig. 8.8.

The confusion matrices show a major improvement in sandpaper identification when the subject wears the feedback device with respect to without feedback. The overall accuracy with the VIBES was 52%, and without the VIBES was 40%. Furthermore, the SUS questionnaire resulted in a positive score of 77.5 (average SUS score of 68 at the 50th percentile).

8.5 Discussion

The Psychophysical Characterization of the VIBES revealed consistent behaviours among able-bodied subjects when discriminating roughness stimuli. JND was equal to 87.30 μm, in line with the results in Libouton et al. (2010) regarding human index fingertip exploration. Indeed, the tactile roughness discrimination threshold at the

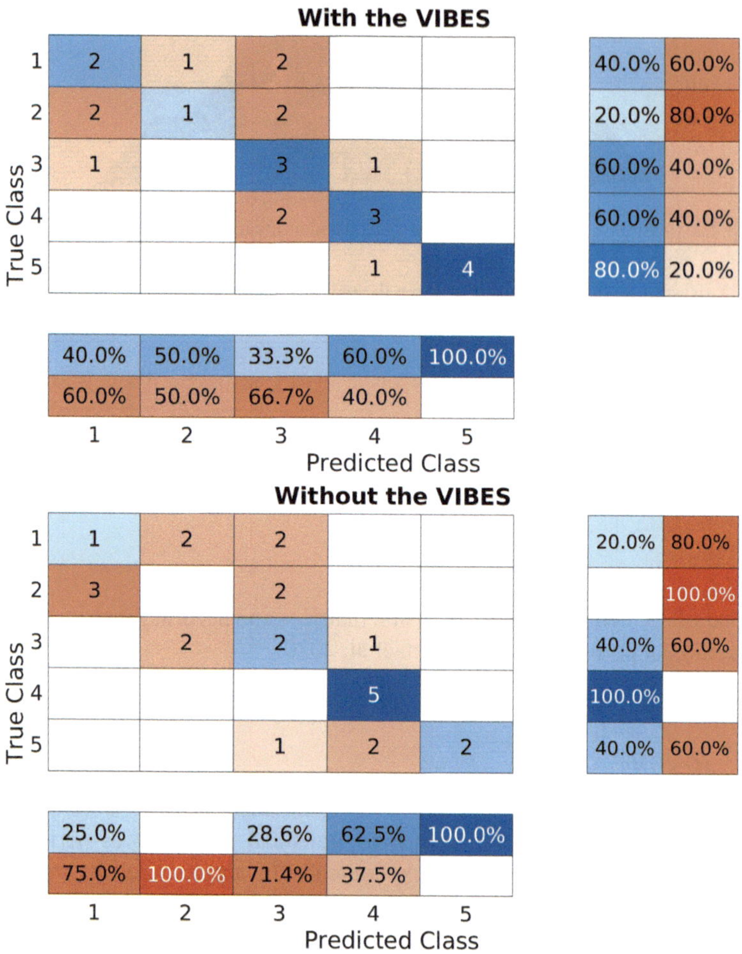

Fig. 8.8 Active identification experiment: confusion matrices with and without the VIBES. Stimuli in descending roughness order, from 1 (P60) to 5 (P100) (see Table 8.1). The row and column summary displays the percentage of correctly classified and incorrectly classified observations for each true or predicted class

index fingertips in humans with active exploration is 15±8.5 μm (smoothed sandpapers) and 44±32.5 μm (roughest sandpapers) (Libouton et al. 2010). Considering that our experiments involved passive exploration of texture through vibrotactile actuators rendering, our results are in accordance with these prior findings. Furthermore, the Psychophysical Characterization Experiment with a prosthetic user yielded positive results, with a Just Noticeable Difference of 44.07 μm. Thus, the HP actuator is able to enable users to perceive and discern roughness with high precision.

The Active Texture Identification Experiment with a prosthetic user yielded favourable outcomes. The participant could better identify the sandpaper presented with the VIBES feedback than without it. Indeed, the accuracy with the vibrotactile feedback was 52% with respect to 40% without the feedback. Although the study results are promising, achieving 100% accuracy was not possible. This can be attributed to the difference in skin sensitivity between the stump and fingers, as there are more receptors on the fingers, which could potentially decrease the subject's discrimination performance. To support this hypothesis, an alternative modality of the experiment could have involved exploring various textures using the residual limb without the prosthetic device. However, we have chosen to recreate a realistic everyday scenario in which the subject, accustomed to wearing a prosthesis, can effectively explore the external environment with the prosthesis. It is also important to note that even able-bodied individuals may not have 100% accuracy in discrimination between similar roughness using their own fingers (Motamedi et al. 2016).

Additionally, the subject's ability to identify the matching sandpaper without feedback was higher than chance, which could be due to her expertise in wearing a myoelectric prosthesis and intrinsic and extrinsic feedback propagation in recognizing different textures. Despite the presence of damping elements in the SoftHand Pro, it is possible that the transmission of vibrations through the socket could have affected the subject's perception. More tests are required to delve further into this aspect.

Based on the subject's comments, the VIBES provided an intuitive experience, and this was further supported by a high score on the SUS questionnaire, which is a promising indication of the VIBES usability and user satisfaction.

It is worth noticing that the results obtained from our study are preliminary, and we acknowledge the need for further research. We aim to conduct a more in-depth physical characterization of the actuators, focusing on analyzing the correlation between the input signal and the resulting signal produced by the actuators. This will provide a deeper understanding of their properties and capabilities. Furthermore, future investigations will comprehensively study how the actuator's position on the stump (e.g. on the hairy or on the glabrous skin) influences how individuals perceive sensory stimuli. Moreover, we will investigate the effect of contact and texture cues even on manual dexterity tasks and with a large sample size to conduct statistical analyses.

It is worth mentioning that the VIBES completes a SoftHand Pro framework, providing the unique advantage of dual haptic feedback options. Indeed, in our previous research (Barontini et al. 2021), we introduced a soft pneumatic integrated system, the WISH device, which effectively conveyed contact and grip force cues to prosthetic users by applying pressure stimuli on their stumps. The WISH device and the VIBES transform the SHP into a highly adaptable prosthesis, empowering the user to select their preferred type of feedback. The comprehensive results derived from the experiments with both able-bodied and limb-loss participants show that the VIBES is a promising solution for transmitting and restoring tactile feedback.

8.6 Conclusion

This study introduces the VIBES device, a vibrotactile integrated feedback system to transmit texture and contact cues to prosthetic users. To evaluate the device's effectiveness, we characterized the system with 15 able-bodied subjects with positive results in terms of texture discrimination performance. From the usability experiments with a prosthetic user, it can be concluded that the VIBES can effectively transmit tactile feedback. The subject was able to use the feedback to improve the success rate in almost all proposed trials. Nonetheless, further research is needed to fully evaluate the device's effectiveness and potential. The positive feedback from the participants, combined with the results of this study, motivate us to conduct further investigations.

Acknowledgements The authors would like to thank Marina Gnocco, Mattia Poggiani and Manuel Barbarossa for their valuable support in the experiments.

References

Akhtar A, Cornman J, Austin J, Bala D (2020) Touch feedback and contact reflexes using the psyonic ability hand. In: MEC20 symposium

Amoruso E, Dowdall L, Kollamkulam MT, Ukaegbu O, Kieliba P, Ng T, Dempsey-Jones H, Clode D, Makin TR (2022) Intrinsic somatosensory feedback supports motor control and learning to operate artificial body parts. J Neural Eng 19(1):016006

Antfolk C, Björkman A, Frank SO, Sebelius F, Lundborg G, Rosen B (2012) Sensory feedback from a prosthetic hand based on air-mediated pressure from the hand to the forearm skin. J Rehabilitation Med: Official J UEMS Eur Board of Phys Rehabilitation Med 44(8):702–707

Balamurugan D, Nakagawa-Silva A, Nguyen H, Low JH, Shallal C, Osborn L, Soares A, Barbosa and Yeow RCH, Thakor N (2019) Texture discrimination using a soft biomimetic finger for prosthetic applications. In: 2019 IEEE 16th international conference on rehabilitation robotics (ICORR). IEEE, pp 380–385

Barontini F, Catalano MG, Grioli G, Bianchi M, Bicchi A (2021) Wearable integrated soft haptics in a prosthetic socket. IEEE Robot Autom Lett 6(2):1785–1792

Bensmaia SJ, Tyler DJ, Micera S (2023) Restoration of sensory information via bionic hands. Nat Biomed Eng 7(4):443–455

Biddiss E, Beaton D, Chau T (2007) Consumer design priorities for upper limb prosthetics. Disability and Rehabilitation: Assistive Technolo 2(6):346–357

Boundy-Singer ZM, Saal HP, Bensmaia SJ (2017) Speed invariance of tactile texture perception. J Neurophysiol 118(4):2371–2377

Choi S, Kuchenbecker KJ (2012) Vibrotactile display: Perception, technology, and applications. In: Proceedings of the IEEE, vol 101, no 9, pp 2093–2104

Cordella F, Ciancio AL, Sacchetti R, Davalli A, Cutti AG, Guglielmelli E, Zollo L (2016) Literature review on needs of upper limb prosthesis users. Front Neurosci 10:209

Della Santina C, Piazza C, Gasparri GM, Bonilla M, Catalano MG, Grioli G, Garabini M, Bicchi A (2017) The quest for natural machine motion: an open platform to fast-prototyping articulated soft robots. IEEE Robot Autom Mag 24(1):48–56

Fani S, Di Blasio K, Bianchi M, Catalano MG, Grioli G, Bicchi A (2019) Relaying the high-frequency contents of tactile feedback to robotic prosthesis users: Design, filtering, implementation, and validation. IEEE Robot Autom Lett 4(2):926–933

Godfrey SB, Zhao KD, Theuer A, Catalano MG, Bianchi M, Breighner R, Bhaskaran D, Lennon R, Grioli G, Santello M, Bicchi A, Andrews K (2018) The SoftHand Pro: Functional evaluation of a novel, flexible, and robust myoelectric prosthesis. PloS One 13(10):e0205653

Grandini M, Bagli E, Visani G (2020) Metrics for multi-class classification: an overview. arXiv preprint arXiv:2008.05756

Jones LA, Tan HZ (2012) Application of psychophysical techniques to haptic research. IEEE Trans Haptics 6(3):268–284

Kim K, Colgate JE (2012) Haptic feedback enhances grip force control of sEMG-controlled prosthetic hands in targeted reinnervation amputees. IEEE Trans Neural Syst Rehabilitation Eng 20(6):798–805

Lewis JR (2018) The system usability scale: past, present, and future. Int J Human-Comput Interact 34(7):577–590

Libouton X, Barbier O, Plaghki L, Thonnard JL (2010) Tactile roughness discrimination threshold is unrelated to tactile spatial acuity. Behav Brain Res 208(2):473–478

Markovic M, Schweisfurth MA, Engels LF, Bentz T, Wüstefeld D, Farina D, Dosen S (2018) The clinical relevance of advanced artificial feedback in the control of a multi-functional myoelectric prosthesis. J Neuroeng Rehabilitation 15:1–15

Moscatelli A, Mezzetti M, Lacquaniti F (2012) Modeling psychophysical data at the population-level: the generalized linear mixed model. J Vis 12(11):26–26

Motamedi MR, Roberge JP, Duchaine V (2016) The Use of vibrotactile feedback to restore texture recognition capabilities, and the effect of subject training. IEEE Trans Neural Syst Rehabilitation Eng 25(8):1230–1239

Raveh E, Portnoy S, Friedman J (2018) Myoelectric prosthesis users improve performance time and accuracy using vibrotactile feedback when visual feedback is disturbed. Arch Phys Med Rehabilitation 99(11):2263–2270

Saal HP, Suresh AK, Solorzano LE, Weber AI, Bensmaia SJ (2018) The effect of contact force on the responses of tactile nerve fibers to scanned textures. Neuroscience 389:99–103

Sakuma M, Kito Y, Tanaka Y, Yoshikawa M, Kawashima N (2020) An electric cosmetic prosthetic hand with vibrotactile sense. In: 2020 42nd annual international conference of the IEEE engineering in medicine and biology society (EMBC). IEEE, pp 4963–4966

See AR, Choco JAG, Chandramohan K (2022) Touch, texture and haptic feedback: a review on how we feel the world around us. Appl Sci 12(9):4686

Shao Y (2022) Compression of dynamic tactile information in the human hand. In: Tactile Sensing (ed) Information, and feedback via wave propagation. Springer International Publishing, Cham, pp 53–78

ten Donkelaar HJ, ten Donkelaar HJ, Broman J, van Domburg P (2020) The somatosensory system. Brain circuitry and its disorders. Clinical neuroanatomy, pp 171–255

Thomas N, Ung G, McGarvey C, Brown JD (2019) Comparison of vibrotactile and joint-torque feedback in a myoelectric upper-limb prosthesis. J NeuroEng Rehabilitation 16:1–18

Chapter 9
Focus on Applications: HRI for Teleoperation

Abstract This work presents the usage of a device for teleoperation tasks, specifically evaluating the impact of wearable haptics for force feedback in teleimpedance control for dual-arm robotic teleoperation. Robotic teleoperation enables humans to safely complete exploratory procedures in remote locations, such as deep-sea exploration or building assessments following natural disasters. Successful task completion in these environments requires meaningful dual-arm robotic coordination and a proper understanding of the environment, capabilities that are inherent to humans via impedance regulation and haptic interactions but challenging to achieve in telerobotic systems. Teleimpedance control has facilitated impedance regulation in such applications, while bilateral teleoperation systems aim to restore haptic sensation to the operator, often at the expense of stability or workspace size. Wearable haptic devices, such as the Clenching Upper-Limb Force Feedback (CUFF), have the potential to inform the operator of key forces during task completion while maintaining stability and transparency. In our study, able-bodied participants completed a peg-in-hole, box placement task, aiming to seat as many boxes as possible within the trial period. Experiments were conducted using both transparent and opaque boxes. With the opaque boxes, participants achieved a higher number of successful placements with haptic feedback, and the experimental trend showed higher mean interaction forces. Results suggest that the provision of wearable haptic feedback may increase operator confidence when visual cues are obscured.

9.1 Introduction

Teleoperation, by definition, is control of a robot located remotely from the operator. To improve operator speed and dexterity, force feedback is often introduced, though it comes with limitations. There are two primary objectives in a bilateral teleoperation scenario. First, the interactions must remain stable. Second, the system should be transparent, so that the operator does not experience the dynamics of the teleoperation system, but rather the remote task. In applications such as building inspections

F. Barontini, *Wearable Haptic Devices for Realistic Scenario Applications*,
Springer Series on Touch and Haptic Systems,
https://doi.org/10.1007/978-3-031-70539-7_9

following natural disasters, robots offer the potential to provide the necessary assessments, while keeping their human operators at a safe distance. Requirements for these robots in building inspections emphasize operability and manipulation, (Liu et al. 2007; Negrello et al. 2018), aspects that heavily depend on feedback to the operator. In undersea applications, the presence of force feedback has been shown to reduce damage to work sites, and to the actual manipulator, though the use of grounded force feedback requires large amounts of training for successful use (Sivčev et al. 2018).

The role of force feedback and its impact on stability has been recognized since the 1960s (Ferrell 1966). The trade-offs between stability and transparency that are experienced in bilateral teleoperation systems have also been well-studied, and have led to a recommendation of providing force feedback to locations on the operator that are separate from where the command input is captured. The addition of kinesthetic force feedback co-located with control input adds latencies to the system, and can result in transmission delay and reflected force feedback (Lawrence 1993; Hannaford 1989). Various approaches have been proposed to address these challenges, including passivity-based approaches (Ryu et al. 2004), model-based feedback via virtual environments (Huijun and Aiguo 2007), or sensory substitution. The sensory substitution method is a promising approach that relies on visual, auditory, or tactile feedback to apprise the user of forces present between the robot and its environment (Schoonmaker and Cao 2006; Prattichizzo et al. 2012), and is an area of active research.

Teleimpedance is an approach to teleoperation where user limb endpoint position and impedance are mapped to the teleoperated robot (Ajoudani et al. 2012). Proposed to overcome stability limitations of bilateral teleoperation, this unilateral approach does not include force feedback to the user, nor does it require a master haptic interface. The teleoperation paradigm maintains desired transparency, and features additional benefits observed in human behaviour. Joint stiffnesses are correlated to the stiffness of the user, introducing advantages such as safety, energy efficiency, and robustness (Santina et al. 2017).

While the teleimpedance framework offers numerous advantages in terms of stability and robustness compared to bilateral teleoperation, the approach still lacks useful haptic feedback to the operator. Wearable haptic devices, such as those proposed by Bianchi et al., offer a unique opportunity to improve the teleimpedance framework (Bianchi 2016). Wearable haptic feedback systems offer a range of actuation modes such as stretch, vibration, and squeeze, and these cues can be used to convey many types of stimuli, including proprioceptive information, navigation cues, and force feedback (Battaglia et al. 2017; Strasnick et al. 2017; Barontini et al. 2024). In the teleimpedance scenario, tactile cues that correlate to some key aspect of the robot's environment can be provided at a location distant from the control input, ensuring stability. In a previous study, this pairing between wearable haptics and teleimpedance was shown to benefit task completion in a single arm telerobotic system, where the haptic device provided feedback on the force of hand closure. When this feedback was provided, performance exceeded that of the open-loop system.

In this work, we explore the benefits of wearable haptic feedback in a dual-arm teleoperation scenario. Here, the CUFF (Clenching Upper-limb Force Feedback) is used to map two force components from the contact interaction at the end effector, as compared to the single grip force of a robotic hand that was provided in prior work (Fani et al. 2018). Two peg-in-hole type experiments are presented, one with visual feedback of the interaction via a transparent box, and the second with an opaque wooden box, obstructing view of the peg and hole. We hypothesize that the addition of interaction force feedback via the CUFF will result in a higher number of successful task completions, and improve force modulation. The dual-arm task design offers the opportunity to convey two aspects related to the interaction, the normal contact force and the shear force due to gravity, leveraging both the squeeze and the twist components of the CUFF. The system demonstrates the potential to leverage the efficiencies of teleimpedance and the intuitive nature of wearable haptic devices in an integrated environment.

9.2 Tele-Robotic Platform

The tele-robotic platform consists of three main parts, the bimanual manipualtor (9.1 left panel), the CUFF wearable haptic feedback system, and the pilot control station (9.1, right panel). A teleimpedance framework is adopted to map the position and stiffness intentions of the pilot in the robot actions. A virtual reality headset, combined with the use of RGB stereo cameras, is used to provide immersive visual feedback to the pilot. A wearable force-feedback haptic device completes the working framework.

1. **Bimanual Manipulator** The bimanual platform consists of two anthropomorphic manipulators (KUKA LWR (Active Variable Impedance)) equipped with a

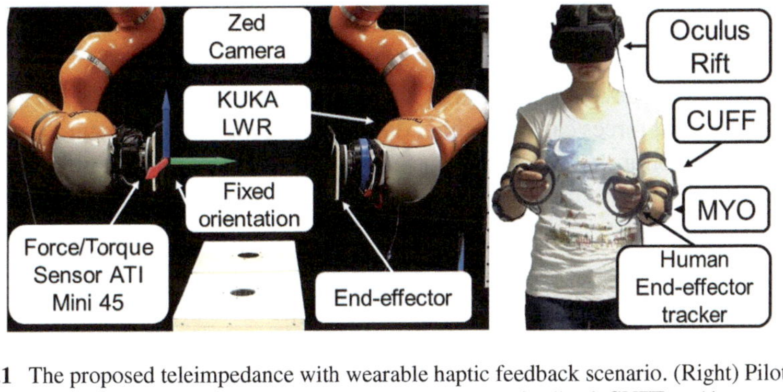

Fig. 9.1 The proposed teleimpedance with wearable haptic feedback scenario. (Right) Pilot interface with Oculus rift for visual feedback, EMG armband, force feedback CUFF, and human motion tracker. (Left) The framework is composed of a stereo camera and 2 KUKA LWR equipped with ATI Mini 45 force/torque sensor

neoprene covered wooden plate End-Effector (EE) (see Fig. 9.1). The wrench at the EE is measured by an ATI Mini 45 force-torque sensor placed between the last robot link and the EE. The KUKA LWR arms are mounted in a bimanual configuration. An RGB stereo camera (Stereolabs Zed Camera Mini[1]) is mounted between the two arms, and is used to retrieve video of the workspace that is streamed to the virtual reality headset.

2. **Teleimpedance** Teleimpedance is a unilateral teleoperation framework in which position and impedance references are sent to the robot to improve the interaction between robot and environment. We adopt the common-mode teleimpedance setup (see Ajoudani et al. 2012). The inverse kinematics of both robotic arms are computed with Priority Inverse Kinematics. The pose of robot's EE is the high priority task given by the pose of pilot's hand, whereas Cartesian position of fourth joint is secondary, to avoid obstruction during the task or camera occlusion. In our setup, the orientation of the EE is fixed (vertical to the floor, facing inward toward the opposite arm (see Fig. 9.1) to limit cognitive loading.
A Cartesian stiffness control law[2] is implemented inside KUKA LWR:

$$\tau_{cmd} = J^T(k_c(X_{FRI} - X_{msr})) + D(d_c) + f_{dynamics}(q, \dot{q}, \ddot{q}) \tag{9.1}$$

where k_c represents the Cartesian stiffness, J is the Jacobian, $D(d_c)$ is a term dependent on the damping value d_c, $f_{dynamics}$ is the dynamic model, and X_{FRI} and X_{msr} are the reference and measured Cartesian positions.

3. **Force Estimation** The forces from the ATI are measured in the EE frame. Since the EE remains stationary relative to the task, the z components, normal to the plate surface, are attributed to the normal interaction forces and are used to calculate the squeeze component on the CUFF (Sect. 9.1 and Fig. 9.1). The force vector is then transformed into the world frame using the known kinematics of the robotic arm, and the downward gravitational component is used to control the twist component on the CUFF.
In a calibration sequence, the CUFF is squeezed around the operators arm until a minimum current value is reached, 45 mA, and repeated for a large maximum current value, 120 mA, is reached. These values are recorded and used to scale the normal force measurements, between 0 and 100 N, to ensure the band on the CUFF does not develop slack during use, and does not squeeze their arm too hard. The tangential forces can be in either direction, and the motors can turn up to 450° in either direction, corresponding to a maximum expected tangential force of 5 N.

[1] https://www.stereolabs.com/.

[2] KUKA FRI 10, Manual for KUKA System Software 5.6 lr.

9.2.1 *Pilot Interface*

Operator movements are obtained with an Oculus Rift[3] (Fig. 9.1). The user's hand pose, given by Oculus joystick, is mapped to the robot EE pose. The reference pose is re-scaled as a function of robot and user arm lengths. Regarding the visual Feedback, the left and right images retrieved from the Zed Camera Mini are streamed to the Virtual Reality Headset with a resolution of 720p and 60 fps. The framework employed for managing impedance includes a wearable device equipped with eight surface electromyographic (EMG) electrodes (MYO Armband[4]). The Cartesian stiffness k_c is derived from EMG on the forearm, calculated as proportional to the average of the readings from the band. This value is the desired stiffness at the EE in Cartesian space, and the necessary joint stiffnesses are then calculated using the current manipulator kinematics.

9.2.1.1 Wearable Haptic Feedback System

The CUFF wearable device is employed for distributed mechano-tactile stimulation of normal and tangential skin forces (CUFF), as described in (Casini et al. 2015). The CUFF consists of two DC motors attached to a band (cuff) worn around the user's arm. When the motors spin in the same direction, the fabric conveys tangential force cues. When the motors spin in opposite directions, they squeeze the belt around the arm, conveying a normal force.

9.3 Methods

9.3.1 *Participants*

Eleven participants, (age 28 ± 2 years, 7 female), all novices to teleoperation control, completed the task with the transparent block. Eleven participants, (age 29 ± 3 years, 4 female, 1 left handed), all with approximately one half-hour of experience manipulating the transparent box (either in pilot experiments or because they participated in the transparent box experiment), completed the task with the opaque box. No participants claimed any physical or cognitive impairment that could interfere with their ability to follow the instructions of the study, nor any pathology that could affect tactile sensation or muscular activity of the upper arm. The methods and procedures described herein were carried out in accordance with the recommendations of the Institutional Review Board of University of Pisa with written informed consent obtained from all users.

---

[3] https://www.oculus.com/.

[4] https://www.myo.com/.

9.3.2 Task Description

To evaluate the benefit of providing wearable haptic feedback in a dual-arm teleimpedance peg-in-hole type task, subjects were asked to manipulate and position a 20 cm cube box such that a peg mounted on the bottom of the box was inserted into a hole in the task environment (1 mm tolerance in diameter). The participant was required teleoperate the dual-arm system to grab the box from the experimenter's hands and seat the box into the base (see Fig. 9.1). The task was completed first with a transparent cube, with clear sides, and then with a wood-faced cube, which obstructed the participants' view of the peg and base (see Fig. 9.2).

9.3.3 Training

For both the transparent and opaque boxes, a ten minute training session was administered before commencing the experiment to familiarize the participant with the hardware, feedback components, and task objectives. The participant was asked to move their arms and explore the reachable workspace of the robot. The box base was moved to ensure box placements were in a natural and comfortable position for the participant. Next, the experimenter pressed on the end effector to create normal and tangential forces, explaining the mapping of the CUFF wearable haptic feedback in the process. The participant was asked to grasp the box with medium density foam

Fig. 9.2 The participants must grasp one of the task boxes (top), either transparent box (left), training box (center) and opaque box (right), and seat it in the base (bottom) in front of the robotic manipulators

attached to the sides (see Fig. 9.2), and with the experimenter still holding the box, modulate the force on the box and note the resulting normal force cues from the CUFF. Next, subjects grasped another box, without foam, while the experimenter pushed on the box from above, demonstrating the tangential feedback cue of the CUFF. Lastly, the experimenter held the box out directly in front of the robotic arms, and the subject grasped the box and seated it in the base. Training concluded when the participant felt comfortable with the task.

9.3.4 Experimental Protocol

Subjects were asked to take a box from the experimenter and place it in the base as many times as possible in a five minute trial. Each time the participant released the box, the experimenter held it up again. In the case of an emergency stop event (due to reaching a torque or force limit), time recording was paused and resumed after resetting the system. The experimenter tapped the end effectors simultaneously at the start of each trial or restart to synchronize the video to the recorded data. Participants completed the task twice, one time with CUFF feedback, and one time without. Half of the subjects complete the trials with haptic feedback first, and the other half completed the trials with no feedback first. Figure 9.3 shows a sequence of the experimental process.

9.4 Data Analysis

Video recordings were used to synchronize data across the various components of the telerobotic platform, and to define the number of task completion attempts, the basis for computing our outcome metrics. Due to complexities in the experimental

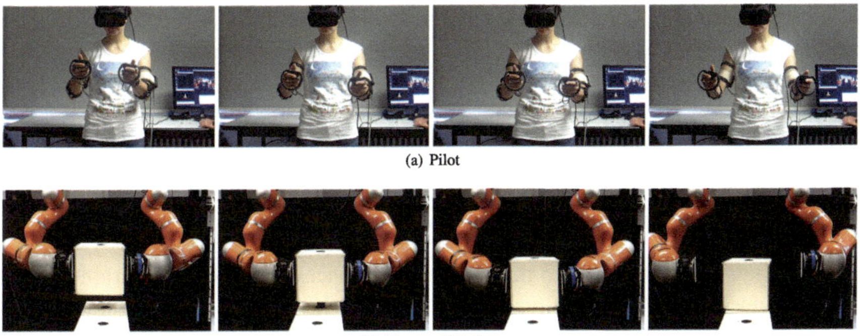

(a) Pilot

(b) Bimanual Platform

Fig. 9.3 Sequences show experimental process: user must grasp box and seat it into base, repeatedly with and without force feedback

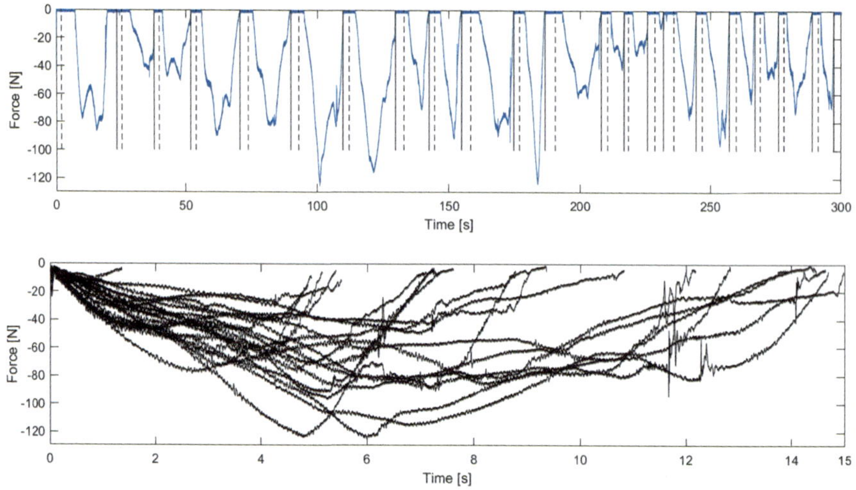

Fig. 9.4 Example force data for an experimental trial, segmented by the beginning (dashed) and end(solid) of each attempt. Below shows the overlaid force data for the segmented attempts

protocol and unsuitable video recordings for some cases, 5 of the subjects in the transparent box task were excluded from the analysis, resulting in 6 participants (age 28 ± 2 years, 4 female) in the subsequent analysis. For the opaque box task, all subjects were included in the analysis.

The video from the stereo camera was recorded along with the force readings from the ATI sensors. After the experiments were completed, the video was analyzed to identify both the time stamps for the beginning and end of each attempt, as shown in Fig. 9.4, and if the subjects successfully completed the task. A successful attempt consisted of the box being seated in the base after release. The time stamps were then superimposed on the force data, and only force metrics from the successful attempts were considered. In each of the two experiments, four metrics were extracted from the attempts, namely the number of attempts to place the box in the base, the number of successful attempts, the mean force during each attempt, and the peak force for each attempt. The values for all attempts were averaged within each trial for each subject.

9.5 Results

Four metrics were evaluated to determine the impact of providing wearable haptic feedback during dual-arm teleimpedance. Each of the four metrics was statistically analyzed to identify differences in each of the three design variables,any any subsequent interactions. Three variables were considered. *Feedback* compares trials with the presence of feedback (On) or without (Off), *Order* considers the order of the On and Off trials, where the On trials are tested first ($FB1$) or second ($FB2$), and *Trial* refers to the two trials ($T1$ and $T2$) completed with each Feedback condition. A $2 \times 2 \times 2$ [Order ($FB1$; $FB2$) $\times$ Feedback (On; Off) $\times$ Trial ($T1$; $T2$)] ANOVA, with repeated-measures on the last two factors, was conducted. Sphericity violations were treated with a Huynh-Feldt adjustment as needed, and simple effects were analyzed using the False Discovery Rate (FDR) procedure.

9.5.1 Number of Attempts

In the transparent box experiment, in the number of attempts per trial, the mean and standard error for the *Feedback* condition was $12.50 \pm 1.80\%$ for On and $11.0 \pm 1.71\%$ for Off, for *Trial*, $11.17 \pm 1.46\%$ for $T1$ and $12.33 \pm 2.03\%$ for $T2$, and for *Order*, $14.25 \pm 2.45\%$ for $FB1$ and $9.25 \pm 2.44\%$ for $FB2$. *Feedback* did trend toward significance when comparing the number of attempts, $F(1, 4) = 6.75$, $p = 0.06$, $\eta^2 = 0.628$, though no other variables had significant effects.

In the opaque box experiment, in the number of attempts per trial, the mean and standard error for the *Feedback* condition was $22.63 \pm 2.57\%$ for On and $20.45 \pm 2.77\%$ for Off, for *Trial*, $20.62 \pm 2.52\%$ for $T1$ and $22.46 \pm 2.77\%$ for $T2$, and for *Order*, $22.13 \pm 3.52\%$ for $FB1$ and $20.95 \pm 3.85\%$ for $FB2$. The opaque box experiment did not have any significant effects for the number of attempts, though there was a trend toward a significant effect between the trials, $F(1, 9) = 4.42$, $p = 0.065$, $\eta^2 = 0.329$. Regardless of the main effects, there were two significant interactions between the *Trial* and *Feedback* presence, $F(1, 9) = 6.72$, $p = 0.03$, $\eta^2 = 0.428$, or between the *Order* and *Feedback*, $F(1, 9) = 6.66$, $p = 0.03$, $\eta^2 = 0.425$. When broken down by *Order*, there was a significant increase in successes with the presence of feedback if the feedback was given second, $F(1, 4) = 21.06$, $p = 0.01$, $\eta^2 = 0.840$, though none of the feedback was given first. When broken down by *Feedback* condition, there was a difference in *Trial* when the CUFF was inactive, $F(1, 9) = 9.373$, $p = 0.014$, $\eta^2 = 0.510$, but not when it was active.

9.5.2 Number of Successes

In the transparent box experiment, the mean and standard error for the number of successes per trial for each of the conditions was as follows: for the *Feedback* condition, $8.17 \pm 1.91\%$ for On and $6.67 \pm 1.81\%$ for Off, for *Trial*, $6.67 \pm 1.44\%$ for $T1$ and $8.17 \pm 2.24\%$ for $T2$, and for *Order*, $9.92 \pm 2.59\%$ for $FB1$ and $4.92 \pm 2.59\%$ for $FB2$. In the transparent box experiment, there was a trend toward significance in the number of successes for *Feedback* ($F(1, 9) = 6.48$, $p = 0.06$, $\eta^2 = 0.618$), though no other variables had significant main effects.

In the opaque box experiment, the mean and standard error for the number of successes per trial for *Feedback* condition was $16.49 \pm 2.04\%$ for On and $13.61 \pm$

1.71% for Off, for *Trial*, 14.44 ± 1.85% for $T1$ and 15.66 ± 1.97% for $T2$, and for *Order*, 14.00 ± 2.41% for $FB1$ and 16.10 ± 2.64% for $FB2$. The sole significant main effect was for *Feedback* ($F(1, 9) = 5.89$, $p = 0.04$, $\eta^2 = 0.396$), showing a higher number of successes with the feedback present, as well as a significant interaction between *Feedback* and *Order* ($F(1, 9) = 5.23$, $p = 0.048$, $\eta^2 = 0.368$). When broken down by *Order*, there was a trend toward a significant increase in successes with the presence of feedback if the feedback is given second ($F(1, 4) = 6.48$, $p = 0.059$, $\eta^2 = 0.631$), though none of the feedback was given first.

9.5.3 Mean Force of Attempts

In the transparent box task, the mean force per attempt trial and standard error for the *Feedback* condition was 57.18 ± 7.88% N for On and 44.07 ± 6.02% N for Off, for the *Trial*, 46.57 ± 5.38% N for $T1$ and 54.68 ± 7.10% N for $T2$, and for *Order*, 48.43 ± 8.02% N for $FB1$ and 52.82 ± 8.02% N for $FB2$. There were no significant main effects or interactions for the mean force across attempts in the transparent box experiment.

In the opaque box task, the mean force per trial and standard error for all attempts for the *Feedback* condition was 65.56 ± 5.53% N for On and 55.43 ± 7.05% N for Off, for *Trial*, 60.55 ± 5.60% N for $T1$ and 60.44 ± 6.64% N for $T2$, and for *Order*, 55.74 ± 8.14% N for $FB1$ and 65.25 ± 8.92% N for $FB2$. The opaque box experiment had a significant effect of *Feedback* ($F(1, 9) = 6.98$, $p = 0.03$, $\eta^2 = 0.437$), and no other variables had significant main effects or interactions.

9.5.4 Peak Force of Attempts

In the transparent box task, the mean peak force per trial and standard error for successful attempts for the *Feedback* condition was 139.51 ± 10.84% N for On and 114.53 ± 5.34% N for Off, for *Trial*, 125.17 ± 8.26% N for $T1$ and 128.87 ± 8.00% N for $T2$, and for *Order*, 108.50 ± 9.78% N for $FB1$ and 145.54 ± 9.78% N for $FB2$. In investigating the main effects, there was a trend toward significance for both *Feedback* ($F(1, 4) = 6.18$, $p = 0.07$, $\eta^2 = 0.607$) and *Order* ($F(1, 4) = 7.17$, $p = 0.055$, $\eta^2 = 0.642$) for the peak force during attempts, but no other variables had significant effects.

In the opaque box experiment, the mean and standard error of the peak force per trial for successful attempts for the *Feedback* condition was 110.00 ± 6.64% N for On and 99.24 ± 9.86% N for Off, for *Trial*, 103.92 ± 7.10% N for $T1$ and 105.32 ± 8.26% N for $T2$, and for *Order*, 97.82 ± 10.16% N for $FB1$ and 111.42 ± 11.13% N for $FB2$. There were no significant main effects or interactions for the opaque box experiment. Results are summarized in Fig. 9.5.

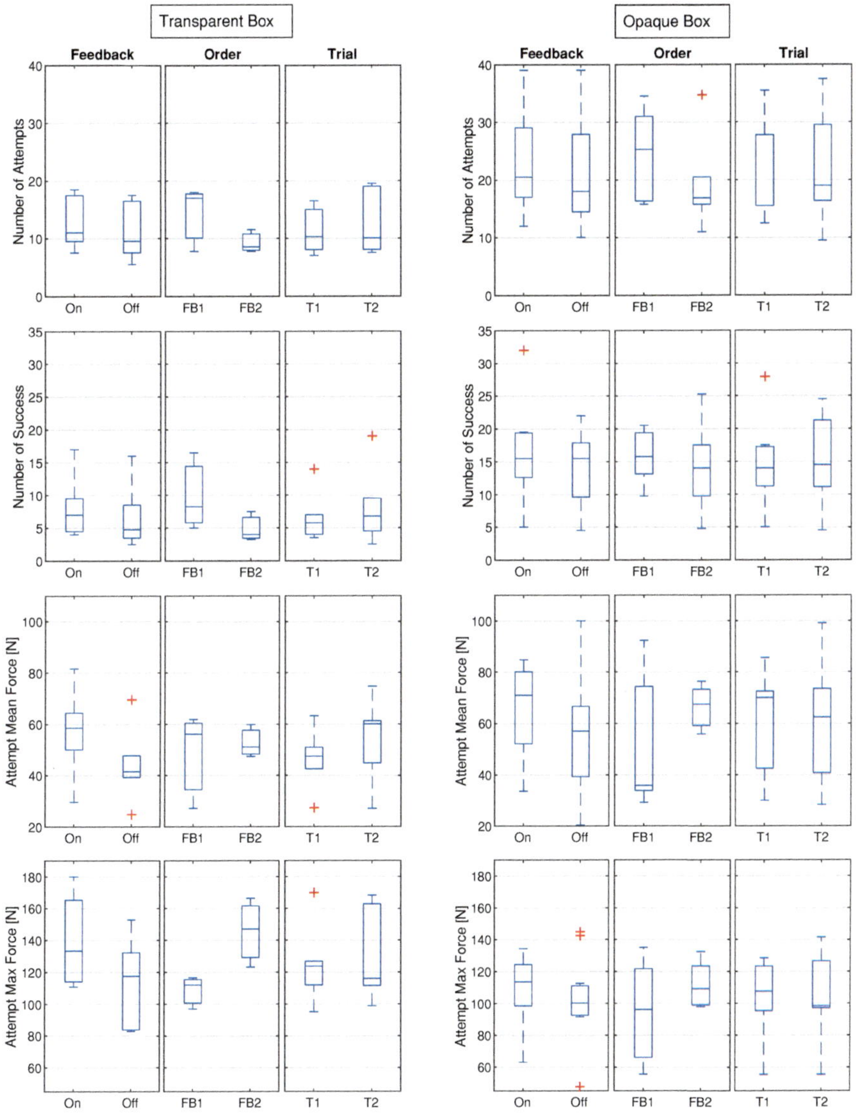

Fig. 9.5 Boxplots summarizing main effects of the number of attempts, successes, and the mean and max force in the attempts for the transparent and opaque box experiments. Feedback compares performance of when the CUFF is active (On) or inactive (Off). Order compares performance of subjects who completed the feedback trials first (FB1) to those that used the CUFF second (FB2). Trial compares the performance between the first (T1) and second (T2) trial

9.6 Discussion

We sought to explore the impact of providing wearable haptic feedback to a human operator when completing a dual-arm teleoperation task in a teleimpedance scenario. Subjects were asked to grasp a customized box, with a peg on the bottom, from the the experimenter's hand, and place it on a base, as many times as possible within the trial time period. Participants completed the task two times in each of two conditions, with and without the CUFF feedback active. The experiment was conducted with both transparent boxes (novice users) and opaque boxes (experienced users).

In the transparent box task, participants were able to locate the hole in the base visually through the transparent box, and did so with and without the assistance of the CUFF. This reliance on visual feedback could be the cause of no statistical differences in the metrics; vision frequently dominates the integrated visual—haptic perception (Warren and Rossano 1991), particularly when the sensory noise is increased (Ernst and Banks 2002). Noting the mean values, the addition of haptic feedback increased mean interaction forces for all attempts and for successful attempts to place the box in the base. Although the difference did not reach statistical significance, the trend suggests that when visual information is limited, participant performance may benefit from the provision of wearable haptic feedback.

In the opaque box task, both the peg and the hole were obstructed from the participant's sight, though sight of the arms and box was maintained. It is not surprising that the number of successful box placements is higher when feedback is available. The total number of attempts with and without feedback is not significantly different; however, both the number of attempts and number of successes demonstrate a significant interaction between feedback condition and order of presentation. Interestingly, there was no difference attributed to the presence of feedback if the CUFF was introduced *before* the no feedback case. That said, there are significantly more attempts made when feedback via the CUFF is introduced after the no feedback trials. Further, the number of successes, while not significant, trends higher in this same scenario. Since this interaction does not exist for the transparent box, one explanation could be that the CUFF provides additional information to the user while they explore and understand the environment, and the repetitiveness of the placement task allows for learning. If the participant experiences CUFF feedback before the no feedback case, they seem to locate the position of the hole, possibly in terms of their body's kinematics, and then continue with relative proficiency after the CUFF feedback is removed. Alternatively, in the absence of CUFF feedback, the hole in the base is harder to locate, and so when feedback is provided later, users benefit and demonstrate improved performance.

Additional, there is a significant interaction between the feedback condition and the number of attempts in the first and second trials for the opaque box. Further analysis shows the difference is only present in the trials without feedback, and we see no difference between the attempts in the first and second trial with the CUFF activated. This result could further corroborate the previously stated hypothesis, where feedback allows the user to quickly accustom themselves to the task. In contrast, when feedback is inactive, there is a learning effect as they try to complete the task without sensory feedback. This trend is not observed in the number of successes. Rather than knowledge of the environment and task space, the results could indicate that users are gaining confidence in their task performance. Future work to investigate these phenomena is warranted.

The introduction of force feedback with the CUFF device is expected to impact the interaction forces between the robot's end effector and the box. For the opaque box, there is significant difference in the mean exerted force within each attempt between the feedback and no feedback cases. Unexpectedly, the forces were higher with the feedback active. Similarly, in the transparent box task, the peak contact forces within each attempt trend toward being higher for trials with force feedback, though no difference is present for the mean force. During both training and the experiment, participants were instructed to maximize the number of successfully seated boxes in the base, with no instruction on force regulation apart from staying below a threshold, which was conveyed through a rapid oscillation of the CUFF only in the feedback condition. Subjects could be using the CUFF as a safety indicator rather than a representation of the robot's applied force on the box. Once the box is grasped, increased forces in the robot-box contact are caused by the participant moving their hands closer together. Since the task objectives centered on speed and accuracy, not minimization of contact forces, a squeeze present in the CUFF informed the user of an adequate grip, and so long as the threshold warning wasn't activated, no other modulation was necessary to accomplish the task. In contrast, when the CUFF was inactive, participants had to be alert not exceed the force threshold since no warning was available. For the transparent box, it is possible that subjects relied on the visual feedback, and when the CUFF provided additional information, subjects had confidence that they would not exceed the threshold, and in turn allowed higher interaction forces between robot EE and box. For the solid box task, the lack of visual feedback could have made participants even more cautious without the CUFF, and more dependent on the feedback as an indicator with feedback active, resulting in larger forces overall. These are complex interactions that require further analysis of the participant's understanding of the system and impression of the feedback.

These findings present several future research avenues. Regarding the confidence of the participants in understanding the environment and how they use feedback, qualitative surveys should be conducted. The impact of learning could be addressed using multiple bases and randomly dictating the target, or by stacking boxes and measuring the time until completion. Lastly, the use of the feedback and the intuitiveness of the force information should be tested by comparing a task focusing on force moderation rather than speed, as they may have conflicting objectives.

9.7 Conclusion

The impact of providing wearable haptic feedback for contact interactions in a peg-in-hole type dual arm teleoperation scenario was explored. The provision of wearable haptic feedback, here via the CUFF, may allow the participant to familiarize themselves with the task and environment more quickly, where without it there were observable learning effects. The addition of wearable haptic feedback resulted in higher contact forces and more successful box placements within the time constraint in the absence of visual feedback.

Acknowledgements The authors gratefully acknowledge Alessandro Settimi for his assistance in implementing the system and Grazia Zambella for their assistance in running experiments.

References

Ajoudani A, Tsagarakis N, Bicchi A (2012) Tele-impedance: Teleoperation with impedance regulation using a body-machine interface. The Int J Robot Res 31(13):1642–1656

Barontini F, Catalano MG, Fani S, Grioli G, Bianchi M, Bicchi A (2024) The cuff, clenching upper-limb force feedback wearable device: design, characterization and validation. In: IEEE Transactions on Haptics

Battaglia E, Clark JP, Bianchi M, Catalano MG, Bicchi A, O'Malley MK (2017) The rice haptic rocker: skin stretch haptic feedback with the Pisa/IIT SoftHand. In: 2017 IEEE World Haptics Conference (WHC). IEEE, pp 7–12

Bianchi M (2016) A fabric-based approach for wearable haptics. Electronics 5(3):44

Casini S, Morvidoni M, Bianchi M, Catalano M, Grioli G, Bicchi A (2015) Design and realization of the cuff-clenching upper-limb force feedback wearable device for distributed mechano-tactile stimulation of normal and tangential skin forces. In: IEEE/RSJ International Conference on Intelligent Robots and Systems (IROS). IEEE, pp 1186–1193

Della Santina C, Bianchi M, Grioli G, Angelini F, Catalano M, Garabini M, Bicchi A (2017) Controlling soft robots: balancing feedback and feedforward elements. IEEE Robot Autom Mag 24(3):75–83

Ernst MO, Banks MS (2002) Humans integrate visual and haptic information in a statistically optimal fashion. Nature 415(6870):429–433

Fani S, Ciotti S, Catalano MG, Grioli G, Tognetti A, Valenza G, Ajoudani A, Bianchi M (2018) Simplifying telerobotics: Wearability and teleimpedance improves human-robot interactions in teleoperation. IEEE Robot Autom Mag 25(1):77–88

Ferrell WR (1966) Delayed force feedback. Human Factors 8(5):449–455

Hannaford B (1989) A design framework for teleoperators with kinesthetic feedback. IEEE Trans Robot Autom 5(4):426–434

Huijun L, Aiguo S (2007) Virtual-environment modeling and correction for force-reflecting teleoperation with time delay. IEEE Trans Indus Electron 54(2):1227–1233

Lawrence DA (1993) Stability and transparency in bilateral teleoperation. IEEE Trans Robot Autom 9(5):624–637

Liu J, Wang Y, Li B, Ma S (2007) Current research, key performances and future development of search and rescue robots. Front Mech Eng China 2:404–416

Negrello F, Settimi A, Caporale D, Lentini G, Poggiani M, Kanoulas D, Muratore L, Luberto E, Santaera G, Ciarleglio L, Ermini L, Pallottino L, Caldwell D, Tsagarakis NG, Bicchi A,

Garabini M, Catalano M (2018) Walk-man humanoid robot: field experiments in a post-earthquake scenario. IEEE Robot Autom Mag 25(3):8–22

Prattichizzo D, Pacchierotti C, Rosati G (2012) Cutaneous force feedback as a sensory subtraction technique in haptics. IEEE Trans Haptics 5(4):289–300

Ryu JH, Kwon DS, Hannaford B (2004) Stability guaranteed control: time domain passivity approach. IEEE Trans Control Syst Technol 12(6):860–868

Schoonmaker RE, Cao CG (2006) Vibrotactile force feedback system for minimally invasive surgical procedures. In: 2006 IEEE International Conference on Systems, Man and Cybernetics, vol 3. IEEE, pp 2464–2469

Sivčev S, Coleman J, Omerdić E, Dooly G, Toal D (2018) Underwater manipulators: a review. Ocean Eng 163:431–450

Strasnick E, Cauchard JR, Landay JA (2017) Brushtouch: exploring an alternative tactile method for wearable haptics. In: Proceedings of the 2017 CHI Conference on Human Factors in Computing Systems, pp 3120–3125

Warren D, Rossano M (1991) The psychology of touch

Chapter 10
Conclusions and Lesson Learned

Abstract This final chapter synthesizes the findings and discussions presented throughout this book, emphasizing the pivotal role of a user-centred approach in developing wearable haptic solutions. The research spans the understanding of neural plasticity and its implications in sensory substitution to the design, development, and validation of novel haptic devices. By involving users from the initial stages of design, this work has led to the creation of innovative devices such as CUFF, WISH, and VIBES, which provide force feedback, and first contact information. These devices have been rigorously tested in diverse applications, including rehabilitation, prosthesis control, and navigation for visually impaired individuals. The findings underscore the importance of user involvement in creating effective and widely accepted haptic feedback systems. This study highlights that relying solely on state-of-the-art technology is insufficient. Instead, an integrated approach that incorporates insights from neuroscience and a thorough understanding of user needs is crucial. Future developments and lessons learned from this research are also discussed, providing a pathway for continued innovation and improvement in the field.

10.1 Conclusion and Future Work

As discussed extensively in this book, all the works presented have been guided by the fundamental idea that user requirements are a crucial aspect of the design process. My research has demonstrated that a User-Centred Approach can significantly contribute to the development of new technologies and methodologies across various fields, including Robotics and Rehabilitation.

The issue of sensory deprivation has been studied not only from an engineering perspective but also from a neuroscientific one. Understanding what happens at the brain level in cases of sensory deprivation, and how the process of neural plasticity is activated, my work has focused on creating haptic devices capable of triggering neural plasticity mechanisms. Special attention has been given to two disabilities intensively explored in the state of the art: blindness and upper limb loss.

F. Barontini, *Wearable Haptic Devices for Realistic Scenario Applications*,
Springer Series on Touch and Haptic Systems,
https://doi.org/10.1007/978-3-031-70539-7_10

Exploring the restoration of missing senses from the user's perspective has inspired the design of a novel wearable feedback device (CUFF), capable of conveying spatial/directional information and force cues. Identifying specific user requirements led to the creation of a prototype that is simple, lightweight, and user-friendly, enhancing daily living tasks and improving users' quality of life. Validating the device through experiments involving a diverse range of participants-from able-bodied individuals to those with limb loss and visual impairments-has allowed for testing in multiple application domains, such as guidance, teleoperation for industrial manipulation, and rehabilitation. The qualitative feedback obtained from these sessions has been instrumental in refining the solution towards a more integrated system.

The next step was to propose two devices (WISH and VIBES), fully integrated within a prosthesis capable of conveying force feedback and first contact cues. These solutions aim to satisfy both the paradigms of modality and somatotopic matching. I believe these solutions have the potential to enhance prosthesis embodiment, not only for expert users but also as a training tool for novice myoelectric prosthesis users, to improve how to better control the grasp of the robotic hand.

The approach and methodologies developed in this research have the potential to influence future advancements in the field of wearable technology. By prioritising user needs, we can create more effective and widely accepted devices that significantly improve the quality of life for individuals with sensory impairments. Future research could explore additional sensory modalities, providing users with multiple sensory information for a more comprehensive experience. Equipping prosthetics with integrated feedback that allows the user to choose whether to activate the feedback only for specific actions could be another path. Another avenue to explore is integrating advanced machine learning techniques for personalised feedback and expanding the application domains to include other forms of rehabilitation and assistive technologies, such as robotic surgery.

10.2 Lesson Learned

Several key developments and lessons have emerged from this research:

- **User Engagement**: The critical role of involving users from the early stages of design cannot be overstated. Users provided invaluable insights that guided the development of more effective and user-friendly devices. Their feedback helped identify practical issues that might not be apparent in a purely engineering-focused approach.
- **Iterative Design and Testing**: The process of iterative design, testing, and refinement proved essential in developing haptic devices that meet user needs. Each round of feedback allowed for improvements that made the devices more intuitive and comfortable for users.

- **Integration of Feedback Systems**: The integration of feedback systems into prosthetic devices, as seen with WISH and VIBES, demonstrated the potential for enhancing user experience and functionality. These systems helped users feel more connected to their prostheses, improving control and reducing cognitive load.
- **Challenges in Sensory Substitution**: One of the significant challenges encountered was ensuring that the sensory substitution provided by the haptic devices was both intuitive and effective. Balancing the complexity of feedback with user interpretability remains a crucial area for ongoing research.

In conclusion, this work represents a significant step forward in advocating for technologies designed for users and by users, starting with their requirements. The insights gained through this user-centred approach highlight the importance of involving end-users from the early stages of design to ensure the development of effective and widely accepted solutions.

GPSR Compliance

The European Union's (EU) General Product Safety Regulation (GPSR) is a set of rules that requires consumer products to be safe and our obligations to ensure this.

If you have any concerns about our products, you can contact us on ProductSafety@springernature.com

In case Publisher is established outside the EU, the EU authorized representative is:

Springer Nature Customer Service Center GmbH
Europaplatz 3
69115 Heidelberg, Germany

Batch number: 10732090

Printed by Printforce, the Netherlands